Textbook of Economic Botany

Textbook of
Economic Botany

Dr. V. Verma, M.Sc., Ph.D.
(Formerly Head of the Department of Botany)
Deshbandhu College
(University of Delhi)

Ane Books Pvt. Ltd.
New Delhi ♦ Chennai

Textbook of Economic Botany

V. Verma

First Edition : 2009
Reprint: 2015, 2016 (Twice), 2017, 2018, 2019, **2020, 2023**

Published by

Ane Books Pvt. Ltd.

4821, Parwana Bhawan, 1st Floor, 24 Ansari Road, Darya Ganj,
New Delhi - 110 002, Tel.: +91(011) 23276843-44, Fax: +91(011) 23276863
e-mail: kapoor@anebooks.com, Website: www.anebooks.com

Branch Office :

- Avantika Niwas, 1st Floor, 19 Doraiswamy Road, T. Nagar,
  Chennai - 600 017, Tel.: +91(044) 28141554, 28141209
  e-mail: anebookschennai@gmail.com, rathinam@anebooks.com

ISBN : 978-81-8052-167-6

Printed at : Repro Knowledgecast Limited, Thane

DEDICATED TO MY BRILLIANT TEACHERS

University of Allahabad

- **Prof. Sri Sri Ranjan**
- **Prof. V S Ramadas**

Ewing Christian College, Allahabad

- **Prof. P S Job**
- **Prof. S N Bhattacharya**
- **Prof. L M Srivastava**

Preface

During the Stone Age, animals were the prime food for survival. With the passage of time, man understood the need for growing of vegetables and plants besides domesticating animals. The nurture of plants led to the practice of agriculture and a strong belief that plants could contribute to the well being of mankind for shelter and clothes.

Since plants do pose unlimited economic importance, a few are popular among human beings. I have dealt with only those plants which are widely used by human beings.

The book is primarily meant for undergraduate students of Indian Universities and has the latest information on the production of cereals and millets of our country. I have tried to describe the economically important plants in a concise and detailed manner which should prove to be useful to the students.

I must acknowledge with gratitude the various authors whose books I have consulted while writing the book. I have taken utmost care to provide authentic illustrations for which I am extremely thankful to Mr. R. K. Majumdar.

Though extreme care has been taken in proof reading, some mistakes might have crept into the book. The teachers and the taught are requested to point out such mistakes to me. It will be highly appreciated.

At the end, I must express my heartfult thanks to Mr. Sunil Saxena and Mr. Jai Raj Kapoor of Ane Books Pvt. Ltd. for the pains they have taken in preparing the book.

I am extremely thankful to my wife Mrs. Santosh Verma for extending a helping hand in the preparation of the book.

V. Verma

Contents

1

Importance of Plants and Plant Products

The importance of plants to the living world hardly needs any elaboration. There are plants and plants the world over and the whole animal kingdom is dependent on them for food. Man's basic necessities of **food**, **shelter** and **clothing** have been met with by plants from times immemorial. In this chapter only a brief outline of the economically important plants or products derived from seeded plants (Spermatophyta) and lower plants is given. **Fossil Fuels.** The main sources of heat and energy for the mankind are **coal** (the carbonized remains of compressed carboniferous forests), **peat** (the partially carbonized remains of recent *Sphagnum* deposits), and **petroleum** (hydrocarbons from ancient plants, mainly diatoms).

Food Plants– The chief sources of human food are:

Cereals– The cereals are sources of carbohydrates. The great ancient civilizations flourished on account of rice (Oriental), wheat (European), and maize (American).

Legumes– Legumes are sources of proteins. They supplement carbohydrates in human diet.

Nuts– A variety of plants bearing nuts, or nut like fruit are rich in fat or protein.

Vegetables– Roots, stem, leaves, flowers and fruits of a number of plants are eaten raw or are cooked as vegetables. Some of them are rich sources of carbohydrates (starch etc.) while others are rich in minerals and vitamins.

Fruits– They are highly nutritious and delicious. The fruits are rich in minerals and vitamins. They are very much liked by human beings. They satisfy their palate and help in the maintenance of good health.

Spices and other aromatic substances– They are used for flavouring food, beverages, medicines, tobacco, and confection and for scenting cosmetics, soap, and incense. They are also used in medicine as carminatives and antiseptics.

Beverages– Tea, coffee, cocoa etc. are used as beverages which have stimulating effects because of the presence in them of certain alkaloids.

Plant Extractives– Sugars, Fats Oils etc. Sugarcane and sugarbeet are of great commercial importance. Sugar, which plays a very important role in our life is extracted from sugarcane and sugarbeet. Seeds and fruits of certain plants yield fatty oils and waxes.

The **vegetable fats** *viz.,* coconut oil, palm oil cocoa butter etc. are solid or semi-solid at ordinary temperatures. The **drying oils** *viz.,* linseed oil, tung oil, walnut oil, soybean oils, perilla oil, etc., harden into elastic films on exposure to air. The **semi-drying oils** *viz.,* cotton seed oil, and sesame oil form soft, solid films only after longer exposure to air. The **non-drying oils** *viz.,* olive oil, castor oil, and groundnut oil do not form solid films on exposure to air. The chief uses of fats and oils are as: cooking medium, massaging oils, and in soaps, candles, linoleum, paints, varnishes, furniture and leather polishes, oil papers, putty, laxatives, lubricants, cosmetics, waterproof fabrics, inks, artificial leather, illuminates and oil cloth, etc. **Waxes** are chemically similar to fats and oils but are harder. They are derived chiefly from leaves (carnauba wax) and fruits (wax myrtle) and are used in the manufacture of candles, wax, varnishes, phonograph records, floor and shoe-polishes etc.

Fibers– Fiber yielding plants are very useful. Man is dependent on fibers for his clothing. The fibre products are textiles, brushes, hats, baskets, chair seats, matting, upholstery and bedding stuffing, packing materials, string, twine, rope, nets, and caulking materials.

Wood Products– The various wood products used by man are for fuel, furniture, house construction, fence posts, mine timbers, poles, pulp-wood, railroad slippers, etc. Hardwoods are obtained from angiosperms and softwoods from gymnosperms.

Wood Derivatives– The substances are derived by chemical and physical alteration of wood. They include charcoal, wood alcohol, acetate of lime, wood tar, wood gases, turpentine, oils; etc.

Tanning Materials– Materials derived from barks (hemlock, oak, mangrove, larch, Norway spruce, birch, willow etc.), woods (chestnut, quebracho), leaves (sumac), roots (palmetto), and fruits (divi-divi) are used in tanning animal hides to convert them into leather.

Dyes and Pigments– They are used in dyeing textiles and leather, paints, varnishes, paper, wood, and ink and for adding colour to beverages, medicines, and foods. They are derived from lichens and woods, leaves, roots, tubers, bark, flowers, and seeds of various plants.

Latex Products– Milky juice of stems and roots of certain plants is called latex. Latex solidifies into elastic or resilient substances on exposure to air or by chemical treatment. Some of the more important latex products are rubber, chicle, gutta-percha, and balata. While rubber is used extensively in the manufacture of tubes and tyres and a variety of useful articles, guttapercha and balata are used for insulating submarine cables, and in the manufacture of machine belting, golf balls, telephone receiver cases and mouth pieces, and surgical splints, etc. Chicle is used in preparing chewing gums.

Gums– Cellulose and other complex carbohydrates on decomposition give a mucilaginous, water-soluble substance called gum. Gums are used in mucilages and glues, paper sizings, calico printing, painting glazes, as stiffeners in ice-creams; meringues and other confections, as an adhesive agent in pills, as a stabilizing agent for insoluble powders in liquid medicines, and as soothing agent in medicines. The chief plant gums are gum arabic, gum tragacanth, gum karaya, peach gum, and cherry gum.

Resins– Resins are water-insoluble, oxidation products of various essential oils, formed in wood and bark of many woody plants. They are widely used in the manufacture of varnishes, lacquers, paper sizings, sealing wax, incense, perfumes, paints, linoleum, ink, and in medicine as healing agents and antiseptics in ointments, etc.

Smoking and Chewing Materials– Tobacco is a well-known smoking material. Tobacco products are widely used the world over for deriving the pleasure of smoking. Opium and marijuana are smoked illegally for their narcotic and hallucinatory effects. Betel leaves and areca nuts are chewed extensively.

Drugs– A number of important drugs are obtained from certain medicinal plants.

Besides the economically important seed bearing plants mentioned above there are some other plants which are also of great economic importance. They are from the lowest plant group—the Thallophyta. The algae are useful as food for human beings, fish and other aquatic animals, as sources of fertilizers, medicinal iodine, agar-agar, as stiffener of confections, ice-creams, shaving creams, mucilages, etc. Fungi are used as food, as sources of vitamins, antibiotics like penicillin, ergotine and other valuable drugs, for manufacturing cheese and other dairy products, alcohols, acetone, enzymes organic acids, etc. Fungi and bacteria decompose the dead bodies and waste products of plants and animals. They maintain the soil fertility and act as agents of sanitation. They help in the production of vinegar, sauerkrant, in the retting of the flax, curing of vanilla pods, in the removal of pulp from coffee and cocoa beans.

2

Food Plants

The great and unique importance of food plants to the mankind is well-known to all of us. The whole existence of the human race is dependent on these plants. The requirements of carbohydrates, proteins, fats, oils, minerals and vitamins by the human body are met with by the food plants. Dry seeds and fruits of cereals, legumes and nuts are the chief sources of food. The roots, tubers, bulbs and leaves of certain plants are important because of their high vitamin contents.

HISTORY OF FOOD PLANTS

All the agricultural crops have been derived from their wild ancestors. Man had domesticated them in ancient times. Since then there plants have undergone minor changes. The last 2000 years have not seen any major breakthrough in the development of new plants.

De Candolle's "L" origin des plantes cultivees," which appeared in **1883** was a classic work on the geographic distribution of useful plants. He used historical writings, archaeological and ethnological data and philology (linguistics) in determining the origins and spread of cultivated plants.

According to **De Candolle** most of the cultivated crops arose from a **single** center but **Vavilov** believed that there are **multiple** centers of origin. According to him 85% of the 640 species have Old World origin and 15% have originated from New World. It is interesting to note that all the centers of origin of crops have extreme variations in their climatic conditions. The mountains of the tropical zones provided ideal habitat for the varieties of the crops to adapt to one or the other prevailing condition.

He classified the useful plants in six classes. Few examples from each of these classes are given below:

A. Old World Species Cultivated for over 4000 years.

Almond	**Cabbage**
Apple	**Cucumber**

Apricot	Date
Banana	Egg plant
Barley	Fig
Flux	Peach
Grape	Pear
Hemp	Rice
Mango	Sorghum
Millet	Tea
Mulberry	Turnip
Olive	Water melon
Onion	Wheat

B. Old World Species Cultivated for over 2000 years and perhaps longer.

Alfalfa	Mustard
Asparagus	Oats
Beet	Orange
Carrot	Pepper
Celery	Plum
Cherry	Poppy
Chestnut	Radish
Cotton	Rye
Lemon	Sugar-cane
Lettuce	Walnut
Lime	

C. Old World Species Cultivated probably for less than 2000 years.

Coffee	Parsley
Gooseberry	Parsnip
Horseradish	Raspberry
Muskmelon	Strawberry

D. New World Species of Ancient cultivation, more than 2000 years.

Cacao	Sweet potato
Kidney bean	Tobacco
Maize	

E. New World Species Cultivated before the time of Columbus.

(Antiquity not known)

Avocado	Potato
Cotton	Pumpkin

Guava **Quinoa**
Peanut **Squash**
Pineapple **Tomato**
Vanilla

F. New World Species Cultivated since the time of Columbus.

Blackberry **Gooseberry**
Black walnut **Pecan**
Blueberry **Plum**
Cinchona **Rubber**
Cranberry **Strawberry**
Dewberry

It is very difficult to trace the wild ancestors and the loci of domestication of the economically important plants, since they were developed in the remote past. An improvement over **De Candolle's** work was that of **Vavilov** (1925–1935).

To De Candolle's findings he added genetical and cytological evidences and took into consideration plant geographical studies and correlation of plant diseases with the host distribution.

Vavilov thought that many primary agricultural crops of today had a diversified rather than a single origin. He suggested two centres of distribution for wheat and three for barley.

According to **Vavilov** the secondary crop plants like rye and oat were originally found as weeds of the primary crops. These weeds dominated over the primary crop plants particulary when conditions were unfavourable for the latter. In course of time the weeds replaced theprimary crop and established themselves as cultivated crop.

According to him neither the humid tropics nor the cold temperate regions have given many food plants. The diversity of environment in mountainous areas favoured the domestication of many food crops.

Vavilov gave a list of six domestication centers which are as follows (Fig. 2.1):

1. **South Western Asia (Afghanistan through Asia Minor).**

   The most important old world center of cultivated plants *e.g.* soft and club wheats, rye,buck wheat, broad beans, peas, carrots, apples, pears, plums, etc.

2. **South Eastern Asia (the mountains of Tibet and China north and east of the Himalayas).**

   The center of origin of plants like oats and barley, soybeans, certain millets, citrus fruits, peaches, and various vegetables.

3. **North-Eastern India-Indo-China (Southeast of the Himalayas).**

   The center of domestication of rice, sugarcane, bananas, bread fruit.

4. **North-Africa (especially Abyssinia).**

   The center of hard wheat, oats, barley, and certain millets.

5. **The Mediterranean Basin.**

   The domestication of cabbage, cotton, olive and fig.

6. **The New World (Peruvian area, and secondarily southern Mexico).**

The centre of domestication of maize, potato, sweet potato, papaya, pineapple and many others.

It is clear from the above classification that whereas rice was the basic food crop in India and China, wheat, barley, rye and oats were the principal crops in the Mediterranean region, Asia minor and southwestern Asia and maize in Mexico and its southward region.

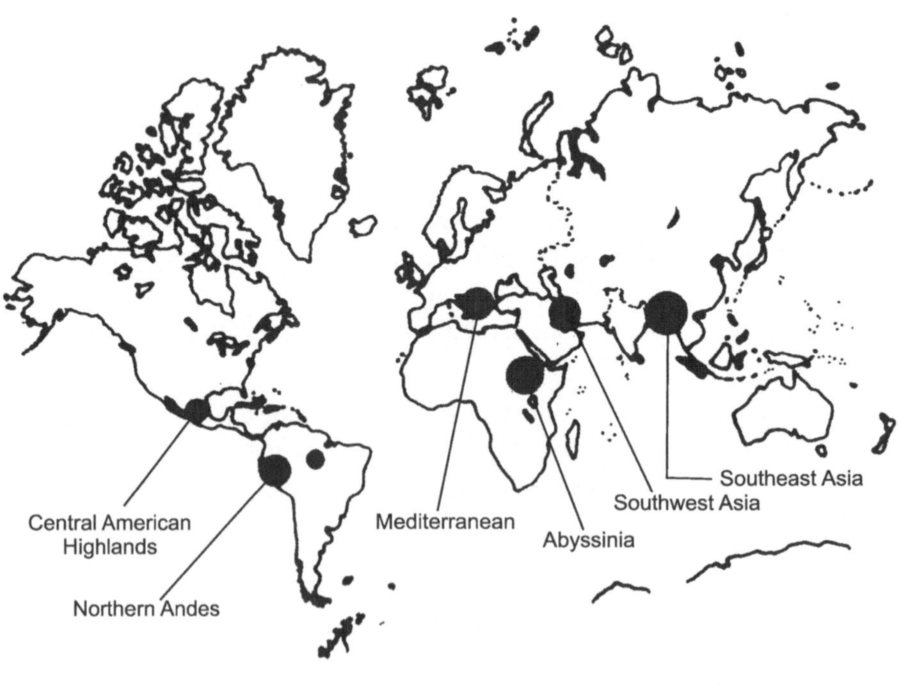

Fig. 2.1. World domestication centers according to Vavilov.

The New and Old World food plants known up to 1492 are given in Table 2.1

TABLE 2.1 New and Old World Food Plants before 1492

New World		Old World		
Maize	Papaya	Wheat	Onion	Apricot
Cassava	Avocado	Rye	Garlic	Peach
Lima bean	Pineapple	Barley	Spinach	Wine grape
Potato	Custard apple	Oats	Eggplant	Pomegranate
Sweet potato	Soursop	Millets	Lettuce	Watermelon
Kidney and other beans	Cherimoya	Sorghum	Endive	Cucumber
Tomato	Guava	Rice	Celery	Olive
Green pepper	Chocolate	Buck Wheat	Asparagus	Fig
Jerusalem	Cashew	Turnip	Peas	Almond
Sunflower	Sapote	Cabbage	Soybean	Quince
Quinoa		Rutabaga	Broad bean	Banana
Squash		Chard	Dasheen	Orange
Pumpkin		Mustard	Yam	Lemon
Peanut		Radish	Apple	Lime
Chayote		Beet	Pear	Date
		Parsnip	Plum	Mango
		Carrot	Cherry	Clover
		Breadfruit	Alfalfa	
		Mangosteen	Bluegrass	

The higher food plants under the following groups are described in the following pages:

(*i*) **Cereals** (*ii*) **Legumes** (*iii*) **Nuts** (*iv*) **Vegetables** and (*v*) **Fruits**.

Zhukovsky (1968) has modified **Vavilov's** concept about the centers of orgin. He has proposed 12 **megagene centers.**

NEW WORLD

Centers of Origin

1. South American Center

(*i*) **Peru, Equador & Bolivia**

1. Potato
2. Tomato
3. Pumpkins
4. Tobacco
5. Egyptian cotton
6. Lima beaned

(*ii*) **Brazil – Paraguay**

1. Cashew nut
2. Cassava
3. Pineapple
4. Groundnut
5. Rubber tree etc.

2. South–Mexico and Central America

1. Maize
2. Sweet potato
3. Red pepper
4. Guava
5. Papaya
6. Squash
7. Bean etc.

OLD WORLD

Centers of Origin

3. (*i*) **The Hindustan Center**

1. Rice
2. Sugarcane
3. Jute
4. Coconut palm
5. Black pepper
6. Brinjal
7. Legumes
8. Mango
9. Orange etc.

(*ii*) **The Indo–Malayan Center**

1. Banana
2. Sugarcane
3. Clove
4. Coconut etc.

4. The Chinese Center

1. Tea
2. Legumes
3. Millet
4. Soyabean
5. Bamboo
6. Cucurbits
7. Onion
8. Cherry
9. Citrus
10. Pear etc

5. The Central Asia Center

1. Wheat
2. Bean
3. Pea
4. Cotton
5. Apple
6. Pear etc.

6. The Near Eastern Center

1. Wheat
2. Rye
3. Pomegranate
4. Walnut etc.

7. The Abyssinian Center

1. Wheat
2. Barley
3. Coffee
4. Castor bean

8. The Mediterranean Center

1. Wheat (Emmer and Durum)
2. Spices
3. Black mustard
4. Cabbage
5. Turnip
6. Garden beet etc.

Later on **Vavilov** (1935) increased the number of centers of origin of crops to **eleven**.

3

Food Plants (Contd.)

THE CEREALS

The cereals are the most important group of food plants. They were the basic food of ancient people and continue to be so in the modern times as well. In fact the very existence of a vast majority of human population depends on the cereals. The economy of a modern nation goes up and down with the rise and fall in the production of cereals.

The modern term '**cereals**' has originated from a festival of ancient Romans. They propitiated the Gods with offerings of wheat and barley (***Cerelia munera***) and called them gifts of **goddess of ceres**.

Cereals appear to have developed in or near mountaineous regions of the subtropics and tropics. The first cereals to be domesticated were the small seeded ***Panicum*** and ***Pennisetum*** which are called 'millets'. Later on the large grained crops seem to have been developed in Ethiopia (Abyssinia), south-east and south-west Asia, southeastern Europe, and the Americas.

It is believed that the long distance of the low latitude mountaineous regions from the northern ice sheets made them suitable for the growth of cereals and with that the growth of civilizations. The fertile mountain valleys, with high rainfall were still better suited for the growth of cereals like wheat and barley. In course of time with the introduction of irrigational systems the cereals spread away from the mountains to the plains. The ancient civilizations of the valleys of the Indus, Tigris, Euphrates and Nile river thus spread and developed to wider areas.

The true cereals are wheat, maize, rice, barley, rye and oats. The pseudocereals are sorghum, millets and buckwheat.

The **major cereals** are as follows:

(i) Wheat (Western Asia and Europe: Temperate climate).

(ii) Maize (premier cereal of the New World, particularly in warm temperature (summer) and subtropical climate).

(iii) Rice (Tropical cereal).

The **minor cereals** are as follows:

(i) Barley
(ii) Oat
(iii) Rye
(iv) Sorghum, and
(v) Millets

Cereals have many advantages as food plants. Cereals contain plenty of carbohydrates and proteins as well as some fats and vitamins. The yield of cereal grains is very large per plant/acre. The grains are compact and dry and so they store well. They are produced in clusters at the tips of the plants. Thus the grains can be easily harvested. The gains can be stored safely because of their low water content.

INDIA'S MAJOR FOOD CROPS

In India prior to Green Revolution in the year 1965–1966 area occupied by food grains was 115 Million Hectares but the total output was only 72 Million Tonnes.

In 2000–2001 the food grains occupied an area of 121 million hectares and the total pruduction increased to 196.81 million tonnes. In 2006–2007 the land area increased to 124 MH and the production to 212 MT. The kharif and rabi crop production are 108.5 MT and 103.5 MT respectively.

The wheat revolution in the sixties raised production of wheat from 11.4 million tonnes in 1966–67 to 74 million tonnes in 2006–2007. It is estimated that the wheat growing area is now more than 281 million hectares. Rice accounts for about 40 per cent of the total production of the food grains in our country. The total production of rice is 91 million tonnes in 2006–2007 whereas it was 30.43 million tonnes in 1966–67 (See Table 3.1).

TABLE 3.1 Showing Production of Important Cereal Crops in Million Tonnes

Year	Rice	Wheat	Jowar	Bajra	Maize
1965–66	30.58	10.39	7.58	3.75	4.82
1975–76	49.45	28.33	9.52	5.72	7.03
1990–1991	74.29	55.14	11.68	6.89	8.96
2000–2001	84.98	69.68	7.53	6.76	12.04
2006–2007	91.05	73.70	7.63	8.28	13.85

In the last forty years the tremendous increase in the agricultural output in our country has been due to mainly the new crop varieties and improved farming practices. Earlier we had numerous varieties of a crop cultivated throughout our country. Today they have been replaced by a smaller number of select high yielding varieties. These new varieties have ushered in the green revolution in our country. In 006 – 2007 these varieties had covered an area of 124 million hectares from less than 100 hectares in 1964–65. It should, however, be clear that the high yielding varieties alone were not responsible for the bumper yields we have had over the years. During this period of forty years large additional areas have been brought under cultivation. Besides increased use of fertilizers, presticides and agricultural machineries have also contributed to the increased agricultural production.

Rice is by far the most important crop of India and 40 per cent of the paddy cultivated area has been brought under the high yielding varieties. One of the dangers of using very few select varieties of rice by farmers (most of the varieties are derivatives of IR-8) is their vulnerability to diseases and pests. This is due to their narrow genetic base since they all have the same dwarfing gene of **dee-geeo-woo-gen**.

A similar danger is faced by wheat, the second major crop of our country. Though several varieties have been recommended for the five wheat zones of our country only few varieties are popular with the farmers. These few varieties have a very narrow genetic base and are hence vulnerable to epidemics.

WHEAT

[Vern. Gehun]

(Triticum spp.)

Origin and History

Wheat is the foremost crop of the world. Wheat grain contains carbohydrates (60–80% starch); protein (8–15%) with all the aminoacids except lysine, tryptophan and methionine; fats (1.5–2%); minerals (1.5–2%) and vitamins such as B complex and vitamin E. These are the essential requirements of the human food. Besides the wheat being highly nutritive, its low water content, easy transport, storage, processing and milling qualities make it the staple food of over one-third of world's population. According to calculations the production of wheat crop is above 450 MT/year.

Wheat is the prime gift of the Old World to the New World as maize is the gift of the New World to the Old World. Wheat is one of the earliest cereals to have been brought under cultivation. Now it is world's most widely cultivated crop, has about 17,000 different varieties and is grown in all the continents except Antarctica.

The origin of wheat is still a subject of speculation. Some archaeological findings of the grains of domesticated wheats date to approximately 7500–6500 B.C. The oldest known remains of wheat discovered are the carbonized grains of the mild, brittle diploid wheat ***T. monococcum*** var. ***boeticum*** ('einkorn' wheats) and of the cultivated, non-brittle var. *monococcum* as well as of the wild tetraploids ('emmer' wheats) from the prehistoric settlement of Tell Mureybit, northern Syria (eighth millenium B.C.), Ali Kosh in Iranian Khuzistan (7500-6750 B.C.) and village of jarmo in eastern Iraq. This area *viz.*, the mountain chains around the **plains of Mesopotamia** and the **Syrian desert** has been termed the **fertile crescent**. It is interesting to note that the wild variety was gradually replaced by the cultivated variety. That the fertile crescent is the site of the origin of domesticated wheat is substantiated by the fact that the wild diploid (einkorn) and the tetraploid (emmer) wheat occur naturally in the south-western Iran, north-western Iraq, south-eastern Turkey. 'Einkorn' is also very common in Transcaucasia, Syria; central and western Turkey and Greece in secondary habitats such as roadsides and field edges. According to many American workers the origin of wheat occurred in **Syria** and **Palestine**. **Vavilov** believed that wheat originated from two centres— the **soft wheat** from **south-western Asia** and **hard wheats** from the **Mediterranean region**. Farming of wheat is believed to have spread to the irrigated settlements in the lowlands of the Mesopotamian plains and the Nile valley and to central and western Europe in the fifth millenium. Subsequently einkorn wheat became widely distributed in Europe and West Asia. Today, however, this wheat is to be found only in some mountaineous parts of Yugoslavia and Turkey.

The wild tetraploid emmer ***(T. turgidum)*** is supposed to have given rise to the most of the cultivated tetraploid and hexaploid wheats. The emmer wheats probably migrated from the mountaineous areas of the fertile crescent to the lowlands of Mesopotamia in the sixth millenium B.C. and to Egypt, the Mediterranean basin, Europe, central Asia and India in the fifth and fourth millenia B.C. Emmer was replaced by the better free threshing tetraploid form, ***T. turgidum*** var. ***durum*** as the principal cereal in the beginning of the first millenium B.C. In modern times emmer is limited to only Ethiopia, Iran, Transcaucasia, eastern Turkey and the Balkans.

Archaeological findings indicate the free threshing hexaploid ***aestivum*** to be present in the **West Asia** from the seventh millenium onwards and the hulled hexaploid ***spelta*** etc. from 2000 B.C. onwards. ***T. aestivum*** var. ***compactum*** has been discovered from Tell. Ramad, Syria (about 7000 B.C.). This wheat is believed to have become an irrigated crop in the plains of Mesopotamia and western Iran in the sixth millenium and in central and western Mediterranean basin in the fifth millenium. Compact firms of hexaploid wheat found way in central and western Europe at the end of the fourth millenium. The hexaploids are believed to have spread to the Indus valley through the highlands of Iran and to the central Asia at the beginning of the third millenium B.C. Records of ancient China show its cultivation there in 2700 B.C.

Triticum aestivum was introduced into Mexico in 1529 and into Australia in 1788 by Spaniards.

Wheat Growing Regions

World—Almost all the wheats are grown in the countries situated in temperate zone (between 30°–50° N and 25°–40° S). U.S.A. is the largest producer of wheat in the world. It accounts for one-fourth of the total world production of wheat. The other wheat producing countries are Russia, Canada, Argentina, Australia, France, China and India. India stands fifth in terms of total yield of wheat in the world.

India– In India, over 28 million hectares of land are sown with wheat and the annual production in currently 73.70 million tons. The main wheat producing states in India are Punjab, Haryana, Uttar Pradesh, Rajasthan and Madhya Pradesh. The wheat growing areas can broadly be divided into five principal zones (Joshi, 1961).

1. **The North Hills Zone**, comprising mountains and submountain regions of Jammu and Kashmir, Himachal Pradesh, Punjab, Uttar Pradesh, Bihar and West Bengal.
2. **The North Plain Zone**, covering the plains of Punjab, Rajasthan (except south-eastern Rajasthan), Delhi Western and Central Uttar Pradesh, and Gwalior region of Madhya Pradesh.
3. **The Eastern Plain Zone**, covering plains of Uttar Pradesh, West Bengal and Orissa.
4. **Central Zone**, comprising Madhya Pradesh, South Rajasthan and Gujarat.
5. **The Peninsular Zone**, including Maharashtra, Andhra Pradesh, Karnataka (Mysore), Nilgiri and Pulney hills of Tamil Nadu.

Wheat occupies 12% of the total area under cereals. It accounts for 75% of the winter cereals. ***Triticum aestivum*** the common bread making wheat is grown in the northern India while *T. durum* is mainly grown on the black soils of central India.

Botanical Description

Wheat like other cereals is a member of the family **Poaceae**. It is an annual with a life span of about 6 months.

The wheat plant usually grows to a height of about 2½ to 5 ft. depending upon its variety and the fertility of the soil (Fig. 3.1). The macaroni ***(T. durum)*** wheats are taller than the common bread wheats.

The height of the plant is also affected by the external factors. The stems or culms of the wheat plant are cylindrical. In most varieties of the bread wheats the stems are hollow except at the nodes where they are solid, but in a few varieties of macaroni wheats the internodes are completely filled with a soft pith. In the early stages, the nodes or joints are very close together but later the internodes elongate until the plant attains its full height. Wheat like other cereals produces branches called "**stools**" or '**tillers**'. The degree of tillering depends on the variety of the wheat plant, fertility of the soil, spacing and seasonal conditions. The wheat plants can be of 'erect', 'creeping' or 'intermediate' types.

Fig.3.1. A wheat plant.

The wheat leaf consists of four parts: *(i)* the blade or lamina *(ii)* the sheath, the part which clasps the stem tightly *(iii)* the ligule or rain guard, a thin membrane growing at the junction of the blade and the sheath and *(iv)* the auricles, which are curved claw like appendages attached to the base of the blade and generally clasping the stem from opposite sides. The roots are fibrous. After the completion of the vegetative period of growth, the wheat plant produces ears or spikes which bear the grains. This phenomenon is known as the '**shooting**' of the ear or '**booting**'. A spike inflorescence of wheat consists of spikelets which are borne on alternate side of rachis, giving it a zigzag appearance. A spikelet may have one to five flowers enclosed in a pair of glumes, which are concave scales or bracts inserted on a short stem known as the rachilla. The rachilla is attached on the rachis by means of cushion-like ends present at the nodes of the latter. Spikes differ greatly in form and degree of compactness. Spikes may be '**awned**' (bearded) or '**awnless**' (beardless). An awn is a tapering projection of the flowering glumes (lemmas) of a spikelet.

A wheat flower consists of three stamens with thread like filaments and rather large anthers. The single ovary has two like filaments and rather large anthers. The single ovary has two feathery stigmas. There are two lodicules which regulate the opening of the flower. Wheats are usually self-pollinated. Cross-pollination and fertilization has been seen in macaroni or durum wheats, more in hot and dry climates than in wet and cold climates.

The wheat grain is a one seeded dry fruit (caryopsis) with the seed coat fused with pericarp. It has a large starchy endosperm and an embryo.

Suitable Plant Type

According to **Asana** (1968) an efficient type of wheat plant for non-irrigated cultivation should have the following characteristics:

Ear	Large number of spikelets or branched ear, so that large number of grains can develop.
Peduncle	The length of ear-bearing stalk (peduncle) should form a substantial proportion of the length of the stem which may be semi-dwarf.
Leaves	About seven on main axis, preferably horizontally disposed (shade the ground better and retain dew longer than those with more vertically disposed leaves).
Root growth	Substantial probably related to time of earing.
Time of earing	Ear must appear at such a time that grains can develop for at least five weeks at a mean maximum temperature of 25°C.

Ecological Factors

Climate— Because of its extraordinary wide spread cultivation, there are wheat plants growing and maturing at all times in a year somewhere in the world. It is, however, not everywhere that climatic conditions are very favourable for the cultivation of wheat crop. The cool winter of the temperate zone is best for the growth of the plant. It is not grown in warm and humid regions because of its susceptibility to diseases.

Wheat is of two type: '**winter wheat**' and '**spring wheat**'. Winter wheats are grown in about two-thirds area of the wheat growing regions. They are usually sown in autumn (October) and are harvested in summer (May-July). They require both **low temperature** as well as **long days**. They germinate at moderate temperatures but begin to flower only when the day lengths are longer in early summer. If they are sown in spring, they behave as biennials. On the other hand the spring wheat is sown in April-May and is harvested in August or September.

In India wheat is grown best in areas with 15–35 rainfall. The soil should be wet to a depth of 3 feet at the time of sowing. The minimum and optimum temperature for the growth of wheat are 3–4°C, and 25°C respectively. High temperature at the pre-harvest time reduces grain yield and warm temperature during early period of growth is also not good.

Irrigation is essential for a better yield of the crop because prolonged drought in unirrigated areas reduces production considerably. Frost at the time of flowering and hailstorms at the time of maturation of the grains are very harmful. The quality of grain produced in relatively dry areas in superior to that of humid or irrigated areas.

Soil—Wheat grows well in fertile, medium to heavy textured soils that are still drained. Highest yields are obtained on the silt and clay loams. The sandy and porous soil is not good for the crop.

Sandy loam soil is good for wheat if barley, mustard or gram are grown along with it. On the basis of soil conditions wheat growing regions of our country may be divided into two, *viz.*

(i) the Indo-Gangetic alluvium of U.P., Bihar, Punjab and Rajasthan, and

(ii) the black soil tracts of south Rajasthan, Madhya Pradesh, Maharashtra, Andhra Pradesh and Karnataka.

Cultivation

The sowing season starts from the second week of October and continues up to the middle of November. In some areas of India (particularly M.P.) it is sown sometimes in September. In Punjab, Delhi and their bordering areas the sowing is usually done in November and sometimes in December depending upon the climatic conditions. Late sowing is not desirable since the crop is likely to be heavily infected with the rust disease.

The field is generally ploughed four to five times before sowing. Proper soil moisture content is essential for getting uniform germination of the seeds.

In India wheat is grown mostly during the **rabi season**, when the amount and intensity of rainfall is low. Tall Indian varieties of wheat require 3–4 irrigations for obtaining maximum yield, with a rainfall amounting to 50–100 mm on medium textured soils of Punjab and Delhi. Coarser soils need more number of irrigations. Sandy loam soils of Rajasthan (Jubner) require 6 to 7 irrigations whereas the light textured soils of Gujrat require as many as 7 to 13 irrigations. Three Mexican varieties of wheat *viz.* Kalyan Sona, Safed Lerma and Sonalika give increased yields with increased wetness. Four irrigations have been found to be optimum.

The seed may be sown by broadcast method or by drilling machines. It is advisable to sow seeds deeper in rough, dry and light soils and shallower in moist and heavy soils. Seed used for sowing should be treated with chemicals to make them disease free and should be highly viable. Weeding is constantly done because wheat is easily harmed by weeds.

Major nutrients required by the wheat plants are nitrogen, phosphorus, potassium and calcium.

Harvesting

Mid January	Karnataka
February	South and Eastern Maharashtra
March	North Maharashtra, Madhya Pradesh, Bihar, Uttar Pradesh and
Rajasthan	
April	Punjab and Delhi
May and June	Mountaineous regions of the North.

Threshing— It involves the separation of the grain from the spike. In our country it is usually done by treading the spikes on ground under the feet of cattle. Winnowing is done with baskets when the husk is removed by the blowing wind. Simple mechanical threshers and winnowers are also used. In western countries threshing machines are generally used. In U.S.A. very complicated machines have been developed, which reap, thresh, clean, winnow, and sift the grains, separate the wheat from the chaff, eliminate other unwanted seeds, sort into grades and bag the grain and also bind the straw. With the help of these machines eight men can harvest 120 acres of wheat crop daily.

Storage— Absolutely dry grains are stored in warehouses which may be buildings with floors and walls of concrete or walls and roofs of iron sheets. Storage bins or pits are also quite common. All such storage godowns should be waterproof. They should be fumigated to keep away the pests.

Cytotaxonomy of Wheat

The genus ***Triticum*** contains several wild and cultivated species. The Russian botanist **N.I. Vavilov** recognized 14 species of wheat. The classification of the species into three groups is based on the original division of the genus by **Schoultze** (1913). This German worker made the anatomical, morphological and chemical properties of the species as the basis of his groupings. The baking properties are affected by the chemical properties of the plant. According to **Sakamura** (1918) the three groups differed in the number of their chromosomes. The first group contained 84 chromosomes in two sets (2 × 7), the second had 28 chromosomes (4 × 7) and the third contained 42 chromosomes (6 × 7).

It is clear, therefore, that different chromosome members play an important role in dividing wheat into three genetically different groups. The 14 chromosome (diploids) species are the basic ones among the three groups. The 28 chromosome (tetraploids) species evolved from a hybrid of diploid wheats and other diploid grasses and their chromosomes had doubled following the process of hybridisation. The tetraploid species are then belived to have crossed with diploid grasses to produce triploids, whose chromosomes doubled and produce the 42-chromosome (hexaploids) species. Since different 'wild grasses' were involved in the evolution of the different species of wheat the resultant species differ not only in the number of chromosomes but also in the genetic contents of these chromosomes. Table 3.2 shows the cultivated wheats and their closely related wild species.

The **diploid Einkorn** species are probably the most ancient. The name Einkorn is derived from the fact that in both the wild and cultivated varieties each spikelet contains only one seed. Wild Einkorn is probably the ancestor of all the other cultivated wheats. In both the varieties the glumes remain clasped around the grain and the flower stalks are fragile. The tendency of the flower stalks to break aids the natural dispersal of seeds but also results in losses during harvesting. The diploid wheats have glumes fitting very closely around the grain and, therefore, they cannot be easily removed. It is, therefore, difficult to use their grains for preparing flour for human consumption. The bread from their flour is dark coloured. The grains are usually used for feeding cattle and horses. 'Einkorn' seems to be simply a domesticated version of 'wild Einkorn' with slightly larger grains and heads which do not fall apart easily. The chromosomes are similar in contents in both the cases which are symbolized AA.

The **tetraploid** species are two in number. They are believed to have evolved from the crossing of diploid 'wild Einkorn' (AA) with a diploid grass ***Aegilops***. The grass involved was probably ***T. speltoides*** which has two sets of chromosomes (BB). The crossing resulted in a hybrid with a constitution of

AB. An accidental doubling of the chromosomes might have given rise to the tetraploid with a constitution of AABB. True Emmer ***(T.turgidum*** var. ***dicoccum)*** resemblas very closely with 'Wind Emmer' ***(T. turgidum*** ver. ***dicoccoides)***. Emmer gives a very good bread. Durum ***(T. turgidum*** var. ***durum)*** is also one of the tetraploid wheats. The glumes of Durum wheat do not enclose the ripe grain firmly and thus it can be threshed free.

The **hexaploid** wheat is the most important one. It is the common bread wheat called ***Triticum aestivum*** (otherwise known as ***T. sativum*** and ***T. vulgare)***. There are six hexaploid sub-species of ***Triticum aestivum***. These are the most recently evolved of the wheats and are the most useful ones, commonly known as the bread-wheats. They have been known as cultivated species. They are the products of hybridization of tetraploid wheats (AABB) with diploid wild grasses (with genetic constitution of DD) with their subsequent doubling of the chromosomes. Thus AABB × DD gives first a triploid ABD, which after the doubling of the chromosomes gives rise to the hexaploid AABBDD, with six sets of seven chromosomes. The grass with which this hybridization took place has been identified as another species of the genus *Aegilops i.e., A. squarrosa,* it is a weed in the wheat fields of Afghanistan.

TABLE 3.2 Classification of cultivated wheats and closely related wild species.

Species	Wild hulled	Hulled	Cultivated free threshing
DIPLOID (2n = 14)			
T. speltoides	all	—	—
T. bicorne	all	—	—
T. logissimum	all	—	—
T. tauschi (A. squarrosa)	all	—	—
T. monococcum	var. *boeticum* (wild einkorn)	var. *monococcum* (cult. einkorn)	— —
TETRAPLOID (2n = 28)			
T. timopheevii	var. araraticum	var. timopheevii	—
T. turgidum	var. dicoccoides (wild emmer)	var. dicoccum (cult. emmer)	var. durum var. turgidum var. polonicum var. carthlicum
HEXAPLOID (2n = 42)			
T. aestivum		var. spelta var. macha var. vavilovii	var. aestivum var. compactum var. sphaerococcum

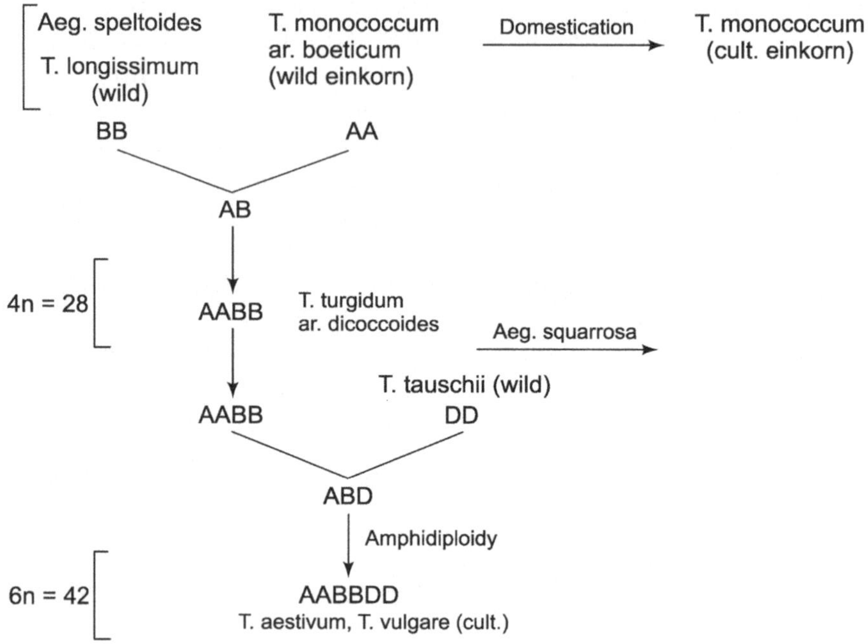

Fig. 3.2. History of diploid, tetraploid and hexaploid cultivated wheats (some of the species of wheat have old names).

The evolution of modern wheat from three parents in ohown in Fig. 3.2.

It is difficult to believe that the ancient people made deliberate attempts to cross different plants for eveloving new species of wheat. They might have, however, played some role in bringing the two parent plants into accidental contact. Natural hybridization was possibly the second step and the selection by man of this useful hybrid was the third step in the **domestication** of the wheat plant.

Unlike the Einkorn and the Emmer wheats the stalks of the inflorescence of the bread wheats are tough and do not shatter at the time of harvesting. The glumes open easily and allow the grain to fall out during the subsequent threshing process.

Important Cultivated Wheat

Bread Wheat ***(Triticum aestivum).*** Hexaploid wheats are the most widely cultivated ones. They are grown in all the parts of the world from the tropics to the sub-arctic, with the greatest concentration is in the temperate grasslands. Their value lies in their productivity and in their being easily threshed, as well as the high content of the **gluten**, the protein that produces a fluffy, leavened bread. The sticky gluten traps the carbon dioxide formed during yeast fermentation and thus facilitates a leavened dough to rise. They are awned (bearded) or awnless (beardless). They include both the spring wheats and the winter wheats. The hard wheats have grains rich in proteins and the soft wheat grains are rich in starch.

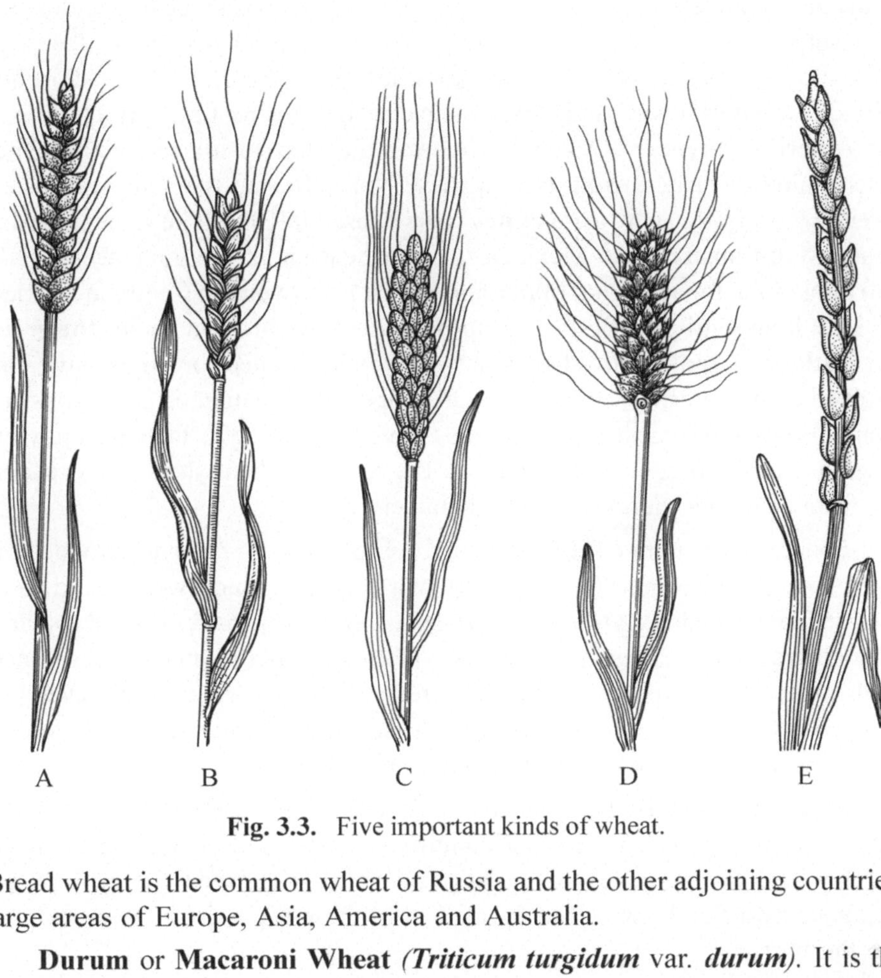

A B C D E

Fig. 3.3. Five important kinds of wheat.

Bread wheat is the common wheat of Russia and the other adjoining countries, large areas of Europe, Asia, America and Australia.

Durum or **Macaroni Wheat** ***(Triticum turgidum*** var. ***durum)***. It is the most common modern tetraploid wheat. Its flour is very good for making macaroni etc., because it contains a high proportion of gluten. It is usually awned. It has spring or semi-winter varieties. It is grown in drier regions *viz.,* throughout the Mediterranean region, Russia, India and other parts of Asia, North and South America.

Some of the important kinds of wheat are shown in Fig. 3.3.

Breeding Wheat

Modern wheat breeders by careful selection and hybridisation seek to achieve the following objectives: (a) better yield of grain, (b) early maturity,

(c) prevention of lodging, (d) drought resistance, (e) disease resistance, (f) quality etc. Lodging of grain laden wheat plants occurs due to rains, windstorms, excessive nitrogenous fertilizers (causes succulence and weakness of the plant), disease etc. Wheat breeding, therefore, aims at breeding for lodging resistant varieties which can withstand heavy doses of fertilizers so that high yield of graincan be obtained. Such varieties must, therefore, have short and stiff straw, short, dwarf and sturdy stems, a vigorous root system, etc. It was the American breeder, **O.Vogel** who was the first to introduce the genes determining dwarfness from the Japanese Norin 10 into the American wheat **Brevor**. A semi-dwarf wheat **Gaines** was released in 1961. The same dwarfing genes were transferred to Mexican spring wheat under the guidance of **N.E. Borlaug** of the International Maize and Wheat Improvement Centre in Mexico. The resulting semi-dwarf (90–120 cm) varieties have brought about '**the green revolution**' in India and other countries. Modern wheat breeders have also evolved early maturing varieties which escape adverse climatic conditions and some diseases. In recent times attempts have been made to increase the yield of grain by effecting cross-pollination. The **hybrid wheat** shows **heterosis** in the form of higher yield and other characters.

Sonora 63, Sonora 64, Mayo 64 and Lerma Rojo 64 A were introduced in India from Mexico in 1964 – 64. The two varieties which were found to be ideal to Indian conditions were Sonora 64 and Lerma Rojo 64 A. The red coloured Mexican varieties had qualities like high yield, excellent tillering, rust resistance. In the next three years new amber coloured varieties of wheat were developed by breeding. The semi–dwarf varieties released in 1967 were **Sharbati Sonora**, **Sonalika**, **Safed Lerma**, **Chotti Lerma** and **PV 18**.

Japanese **Norin 10** dwarf wheats were discovered by **Dr. Salmon** in 1948. The three dwarfing genes have a cumulative effect and depending upon the number of dwarfing genes the varieties of wheat are termed one gene dwarf, two gene dwarf and three gene dwarf. Out of these triple dwarf gene varieties are the best.

Dr. Swaminathan mutated the red–kernelled grains of **Sonora 64** by giving different dosages of gamma and UV rays to obtain **amber** coloured mutants of wheat. Sonora 64 was selected for its attractive amber grain colour. Sharbati Sonora is superior to all other varieties in protein potential (as high as 20%), fairly high lysine content (upto 3.17%), and higher biological and net protein utilization values. The excellent nutritional superiority of Sharbati Sonora over Sonora 64 and some improved Mexican, German and Danish wheats can be further seen from the results of **Prof. Bjorn O. Eggum,** the well known nutritionist (Denmark). Some of the results reported by him regarding biological value and net protein utilization are as follows:

Sharbati Sonora 60.5 and 53.9: **Sonora** 64–52.0 and 48.4; **Super X** (Mexican) 52.0 and 49.1: **Inia** (Mexican) 49.4 and 44.4 ; **Jarral** (Mexican) 49.4 and 45.8; **Colibri** (German) 55.0 and 50.3; **Average Danish** 59.0 and 52.9 etc.

Sharbati Sonora has also very good bread making quality. Another high yielding variety which ushered in the wheat revolution in India is **Kalyan Sona**. The other high yielding varieties of wheat are **Sonalika**, **Hira**, **Moti**, **Shera** and several others.

Mention may be made here of the evolution of a new crop called ***Triticale*** by making intergenere hybridization between wheat ***(Triticum)*** and rye ***(Secale)***. It resembles wheat in many ways and has greater vigour than wheat.

Some High Yielding Wheat Varieties

1. Sonalika (HD—1553) released in 1967.

This is an early maturing variety. It is a single gene dwarf wheat with attractive grains, resembling good quality desi wheats.

It is suitable for timely as well as late sowings in U.P., Haryana, Delhi, Rajasthan, M.P., Maharashtra, A.P., Tamil Nadu and Karnataka.

2. Kalyan Sona (HD—1593) released in 1967.

It is a medium late maturing variety. It is high yielding wheat with widest adaptability.

It has been grown in Jammu and Kashmir, Punjab, Haryana, Delhi, U.P., Rajasthan, M.P., Bihar, Orissa, West Bengal and Maharashtra.

3. Sharbati Sonora released in 1967.

Sharbati Sonora is an amber mutant of Sonora-64 with early maturity and synchronous habit of tillering. It is grown in all wheat growing regions.

4. Shera (HD—1925) released in 1974.

It is a double dwarf wheat with very good bold amber grains. It is resistant to lodging, shattering and black rust in Central and Western zones of our country.

5. Raj—911 released in 1974.

It is a 2-gene dwarf durum. It is high yielding as well as resistant to rust. It is suitable for the central wheat tract.

6. Malvika (HD—1502).

It is a triple dwarf durum. It is suited for peninsular wheat tract.

7. W1711, **Up368** and **HD2177** are recommended for cultivation under timely sown irrigated high fertility conditions of Punjab, Jammu area, Haryana, Delhi, Western Uttar Pradesh and Rajasthan (except Kota and Udaipur Divisions). These varieties are better than Kalyan Sona and Sonalika in yield and rust-resistance.

8. **WL410** and **C 306** are good for cultivation under low fertility rain-fed conditions of north western India.

9. **UP 115** and **HP 1209** are good for irrigated high fertility and late sown conditions of Bihar, eastern Uttar Pradesh, West Bengal, Assam, Orissa and other eastern states.

10. High yielding and disease resistant **CC 464** and **HD 2189** are good for peninsular India comprising Maharashtra, Karnataka, Andhra Pradesh and the plains of Tamil Nadu.

11. HD (Hybrid Delhi) 2204. It is for the first time that a variety of wheat combines high yield and disease resistance. It has been recommended for large-scale cultivation under high fertility irrigated conditions in the north-western plain zone comprising the country's wheat bowl areas of Punjab, Haryana, Rajasthan, Uttar Pradesh, Delhi and Jammu and Kashmir.

12. IWP-72. It has been recommended for the above mentioned zone of HD-2204 for rain-fed cultivation.

13. HW-657. This highly disease-resistant variety has been recommended for rain-fed cultivation in the peninsular zone states of Maharashtra, Karnataka and Andhra Pradesh.

14. K-7410 and **HUW-12.** These two varieties have been recommended for irrigated areas in the north-eastern plains.

15. VL-421 (Vivekanand Laboratory, Almora). This varieity has been recommended for the north hill zone.

16. The Punjab Agricultural University released a new variety of wheat called **WL-711**.

A Brief account of some of the high yielding varieties of wheat is given below:

Variety	Description	Areas of Adaptability
Kalyansona	A double dwarf variety released jointly by **IARI** and the Punjab Agricultural University. Most widely grown in India presently. Suitable for cultivation under both normal and late plantings as well as high and low fertility conditions and irrigated and rainfed areas. Yields range from 60–70 quintals/ha. Grains are amber, medium and lustrous. Highly resistant to loose smut and hill bunt diseases.	Suitable for cultivation throughout India.

(*Contd.*)

Variety	Description	Areas of Adaptability
Sonalika	A single dwarf variety next in popularity to Kalyansona in India. Grains are bold, hard, lustrous and very attractive. Sonalika has high field resistance to black and brown rusts. It is suitable for cultivation under both normal and late plantings but particularly suitable for the later category of conditions. Yield potential is from 50–65 quintals/ha.	Suitable for cultivation throughout India.
Sharbati Sonora	An amber grained double dwarf variety produced through irradiation of the red seeded Sonara-64. It has high resistance to black rust. Grains are amber, hard, lustrous and of medium size. Protein content high (upto 16%). One of the best wheats today in India for breadmaking purposes. Yield potential 50–65 quintals/ha.	Late planting in Punjab, U.P., Rajasthan, and normal plantings in M.P., Bihar, W. Bengal, Gujarat, Maharashtra, A.P. and Tamil Nadu.
Safed Lerma	A single dwarf wheat with white, soft grain, highly resistant to black and yellow rusts. Suitable for cultivation under normal sown and high fertility conditions. Yield potential 40–50 quintals/ha.	Peninsular India Particularly Nilgiris and palni hills.
Choti Lerma	A double dwarf wheat with high resistance to black and yellow rusts. It has white soft grains, yield potential is 40–50 quintals/ha.	For the hills of Northernand Southern India, and the states of Peninsular India.
Pusa Lerma	This is an amber grained mutant of Lerma Roja-64A. It is a single dwarf wheat with high resistance to black and yellow rusts, yield potential is 40–50 quintals/ha.	For Karnataka, A.P. and Tamil Nadu.
Hira	A triple dwarf variety with high resistance to yellow rust. Grains amber, hard and long. Suitable for high fertility irrigated conditions. Yield potential is 70.	Western Plain Zone consisting of States of U.P., Haryana, Rajasthan and Punjab.
Moti	A triple dwarf variety with high yield potential upto 75 quintals/ha. Suitable for	Western U.P. where yellow rust is not a

(Contd.)

Variety	Description	Areas of Adaptability
	high fertility irrigated and normal sown conditions. Grains are amber, medium size, lustrous and attractive.	problem.
Shera	A dwarf variety with amber, medium, hard lustrous grains. Performs well under both normal and late sown conditions. Resistant to black and yellow rusts and mederately so to brown rust. Yield potentials is from 45–60 quintals/ha.	Suitable for high fertility, late sown conditions of Central India and normal sown high fertility conditions of the North Western plains zone.
HD 1982	A double dwarf variety with very good amber, bold, hard lustrous grains. Performs well under normal and late sown high, fertility conditions. Highly resistant to brown rust. Yield potential is 42–50 quintals/ha.	Suitable for normal sown, late sown high fertility and medium fertility conditions in the Eastern Zone.
HB 102-41	A single dwarf variety with amber, bold grains. Performs well under high fertility conditions in the Hills zone. Resistant to black and yellow rusts. Yield potential is 45–50 quintals/ha.	Suitable for high fertility conditions of the Hilly areas.
HS 1097-17	A single dwarf variety with amber, bold, hard grains, performs well under high and medium fertility conditions in the Hill zone. Yield potential 42–50 quintals/ha. Resistant to black and yellow rust and moderately to brown.	Suitable for high fertility conditions of the Hills zone.
HD 4502	A stiff straw double dwarf durum wheat with bold, amber, hard grains. Performs well under high fertility conditions. Resistant to black and brown rusts, capable of yielding 35–40 quintals/ha.	Suitable for high fertility conditions in the Peninsular Zone.
Meghdoot	A durum variety with excellent bold, amber, hard, lustrous grains. Performs well under rainfed conditions. Fairly resistant to black and brown rusts. Yield potential 12–15 quintals/ha under rain-fed conditions.	Suitable for dry areas of Central zone.
Arjun	Medium late in maturity; resistant to all the three rusts; suited to timely sown, high fertility conditions. Average grain yield is 48 quintals/ha.	Suitable for dry areas of Central zone.

(*Contd.*)

Variety	Description	Areas of Adaptability
Janak	Medium late in maturity; grains bold; resistant to brown rust; suitable for timely as well as late sowing; suitable for medium as well as high fertility conditions. Average grain yield is 41 quintals/ha.	Eastern U.P., Bihar, W.B., Orisa, Assam, Nepal and Bangla Desh.
Pratap	Medium early in maturity; grains bold; resistant to brown rust; suitable for timely as well as late slowing; suitable for low as well as medium fertility condition. Yield 32 quintals/ha.	Punjab, Haryana, Western U.P., and Northern Rajasthan.
Girija	Medium early in maturity; resistant to brown and yellow rusts. Grain yield 27 quintals/ha.	Higher hills of Northern India.
Shailaja	Medium early in maturity; grains amber and bold; resistant to yellow and brown rusts. Grain yield 37 quintals/ha.	Lower hills of H.P.
HD 2122	Early in maturity; grains excellent; resistant to all the three rusts; suitable for timely as well as late sown, high fertility conditions. Grain yield 50 quintals/ha.	Punjab, Haryana, U.P., Northern Rajasthan.
HD 2177	Medium late in maturity; grains amber and bold; resistant to all the three rusts; suitable for early and timely sown, high fertility conditions. Grain yield 42 quintals/ha.	Haryana, Punjab, Western U.P., North Rajasthan.
HD 2189	Medium late in maturity; very good grain quality; tolerant to black and brown rusts, suitable for timely sown, high fertility conditions. Average grain yield 43 quintals/ha.	Maharashtra, Karnataka, A.P., and Tamil Nadu.
HD 2204	Medium late in maturity; good grain quality; resistance to all the three rusts; suitable for timely sown, high fertility conditions. Average grain yield 44 quintals/ha.	Haryana, Punjab, Western U.P., Northern Rajasthan.
HD 2160	Medium early in maturity; grains amber and bold; resistant to all the three rusts; suitable for timely as well as late sown high fertiltiy conditions. Average grain yield 45 quintals/ha.	Haryana, Punjab, Western U.P. Northern Rajasthan.

(*Contd.*)

Variety	Description	Areas of Adaptability
HD 4530 (Durum)	Medium late in maturity; grains amber, long and hard; resistant to all the three rusts; suitable for timely sown, high fertility conditions. Average grain yield 41 quintals/ha.	H.P., Gujarat, Southern Rajasthan, Southern U.P.
Nilgiri	Medium early in maturity; very good grain quality; resistant to all the three rusts. Average grain yield 35 quintals/ha.	Nilgiri and Palni hills or Tamil Nadu.

The Milling of Wheat

The milling of wheat for flour involves a preliminary **cleaning** and **scouring** to remove all foreign seeds, dust sticks, straw and pieces of bran (pericarp and aleurone layer). The grains are then moistened or '**tempered**', a process that 'toughens' the bran and prevents its fragmentation. Tempering prepares the grain into the best conditions for milling. The grains are then subjected to breaking, grinding and rolling in a modern roller flour mill. The grains are run between a series (five sets of rollers moving at different speeds) of corrugated rollers to break or crush the grains. The bran and flattened embryos are screened off from the endosperm in shokes, for use in cattle feed, the endosperm fragments or middlings are then pulverized between smooth rollers to become flour. The flour is finally allowed to be sieved by a silk cloth with 12,000 meshes per square inch. The flour is divided into atta, suji and maida depending upon the size of the particles. It should be noted that hard wheats generally average 13 to 16 per cent protein, and soft wheat 8 to 11 per cent.

Uses of Wheat

1. Wheat flour is suitable for bread making because of its two binding proteins, glutenin and gliaden. "Stronger' wheat flours are those that contain higher percentage of protein resulting in a stickier or more tenacious dough for baking.
2. Soft wheat is preferred for preparing biscuits, cakes, pastries. Hard wheat is good for making breads. Very hard wheat (durum wheat) is ideal for the preparation of noodles, macaroni, pasta etc.
3. Wheat bran is recommended to be consumed along with wheat flour because of its high protein content and also for its richness in vitamins. The bran is not digested within the human intestine and therefore, prevents constipation.
4. Wheat straw is used for making insulated building board.
5. Wheat straw is also used for making rough paper.

6. It is useful as compost.
7. It is used in the preparation of beer and other alcoholic beverages.
8. The bran is a high grade animal feed.
9. Wheat straws are used fodder.
10. Dry wheat straws are used in packing and stuffing. They are also used in the manufacture of straw carpets, straw hats and baskets, etc.

Diseases of Wheat

The wheat crop is subjected to a number of bacterial and fungal disease. Common diseases of wheat are as follows:

1. Rust disease caused by a number of species of ***Puccinia.***
   (*i*) ***P. graminis tritici*** – Black rust.
   (*ii*) ***P. striiformis – (P. glumarum)*** – yellow rust,
   (*iii*) ***P recondita (P. triticina)*** – orange rust.
2. Smut disease caused by ***Ustilago tritici*** – loose smut.
3. Bunt or stinking smut caused by ***Tilletia foetida, T. caries****, etc.*

MAIZE OR CORN

(Vern. Makka)

(Zea mays)

Origin and History

Maize or Indian corn has been the chief contribution of the New World to the cereals of the world. It ranks as one of the four principal crops of the world. While in countries like India and Argentina it is grown for food, it is grown as feed for livestock in other countries.

Maize is believed to have originated in the obscure past. It is very difficult to determine the ancestry of the plant because it does not resemble with any known wild plant. It is believed to have developed from some pod-like ancestors which have been discovered in some of the earliest archaeological deposits. Popcorn has also been found in ancient graves. From genetical evidences it is believed that the ancient maize probably hybridized several times with maize relatives such as ***Tripsacum,*** and ***Euchlaena*** and perhaps with unknown strains of maize now extinct. ***Teosinte (Euchlaena mexicana)*** of **Gautemala** and **Mexico** is believed to be one of the possible ancestors of the modern maize plant. According to **Vavilov** the area comprising Guatemala and Mexico was probably the center of origin for the wild relatives of maize are present here. In mid-1950s archaeological findings of pollen grains of maize in Mexico City have been found to be 60,000 to 80,000 years old. Small cobs of 'wild maize' have been dated 5000 B.C. in the archaeological findings of Tehuacan, Mexico. The findings in the dry coastal areas of Peru, South America have been dated 1000 B.C. According to many authorities there are not one but two centers of origin of the corn plant: (*i*) **Andean centre** in northwestern **South America** and (*ii*) **northern Central America** and **Mexico**. The two areas show great diversity in the maize plants. The Portugese explorers are believed to have introduced the crop into India, China, Malaya and East Indies (now Indonesia etc.) during the sixteenth and seventeenth centuries.

Popcorn with the smallest kernel size is believed to be the first domesticated corn. The commercially important dent types are believed to have spread to the tropics after 1500 A.D. from Mexico and lowland Central America. They also spread to the southern United States. Crosses between these dents and the local indigenous flints of U.S.A. gave rise to the famous Corn Belt dents. The Corn Belt dents are now widely cultivated in southern Europe.

The Cuban and Argentine yellow to orange flints (Catetos) spread to the Atlantic coast after 1600 A.D. (Fig. 3.4). While they are supposed to have given rise to the true flints of the tropics and the subtropics, the northern American flints have given rise to the temperate flints. The origin of flints is quite obscure. The coastal tropical semi-flints and similar **Tusons** of the Carribean and the northern end of South America are believed to have arisen as a result of a cross between the Mexican dents and the indigenous flints of the Carribean.

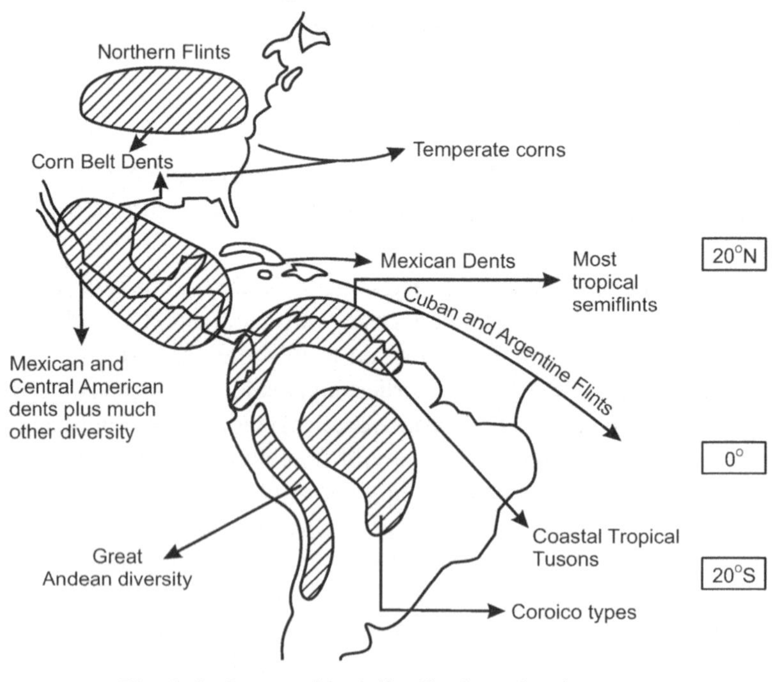

Fig. 3.4. Geographical distribution of maize crop.

A very important centre of variability in maize is to be found in Guatemala. Several varieties of long-eared flint corns occur in this region.

Maize Producing Regions

World—More than half the total world corn crop is grown in Corn Belt region of the United States of America. The average area under this crop in the world is 223 million acres with a grain production of six billion bushels (566 million short tons) and the average acre yield being 25.5 bushels. Leading countries in corn production are United States of America, Brazil, Russia, Mexico, Italy, Yugoslavia, Rumania, Hungary, Argentina, Union of South Africa, India and Indonesia.

America exports only 5 per cent of its crop while the second largest exporter is Argentina. South-eastern Europe and South Africa also account for a larger export of maize.

India—Maize is chiefly grown in Uttar Pradesh, Bihar, Punjab, Madhya Pradesh, Maharashtra, Rajasthan and Orissa. In our country hybrid maize has become an important food crop. Maize occupies more than 10 million acres of land with an average production of over 7.5 million tons.

Botanical Description

Maize is a member of the family of grasses the **Poaceae**. It is a tall herbaceous annual. Its height is very variable. The average height varies from 5 and 10 feet. In some cases the height may be more than 10 feet. The stem is made up of a succession of nodes and internodes (Fig. 3.5A and B). Unlike the other cereals the stem has a solid pith. The plant is fixed into the soil by means of prop roots which arise from the basal nodes of the stem. The inflorescence is of two kinds: (i) The pistillate inflorescence or ears are borne at the side of the plant and are covered by protective leaves or husks, *(ii)* A single staminate inflorescence or tassel develops at the top of the stalk (Fig. 3.6). The pistillate inflorescence is borne on a short branch called the "shank". The pistillate spikelets are 1 to 3 and are borne on a thick axis called '**cob**'. Each ear has 8–24 rows of about 50 kernels each. Each spikelet has two flowers, one of which is abortive. The fertile flower has glumes and palea and a carpel. The styles are very long and emerge through the husk. They are collectively known as "silk". Maize is ordinarily cross-pollinated, but self- pollination is also quite frequent. Pollination may be the agency of wind or gravity.

The grain of maize is a one seeded caryopsis fruit. It has a starchy endosperm and an embryo separated from one another by means of an aleurone layer (Fig. 3.7). The size of the grain (kernel) is proportional to the size of the endosperm. The endosperm is variously coloured (white, yellow, orange). The outer layer of the aleurone may be colourless, red, purple, brown or lemon-yellow. The pericarp is colourless, red, yellow, brown, purple or pink. The pericarp and/or aleurone may be of different patterns (stippled, dotted, speckled, strippled, streaked, etc.)

Fig. 3.5. A. A maize crop (Photo by N. Singh).

Fig. 3.5. B A photograph of part of a plant of maize.

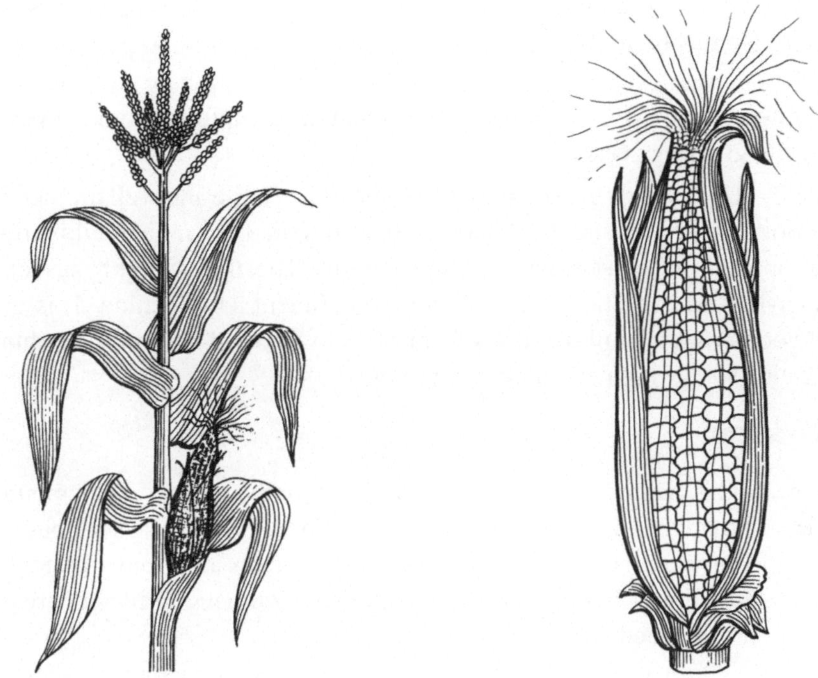

Fig. 3.6. Maize plant. A. A portion of the plant showing staminate and pistillate inflorescences; B. A cob of maize with grains.

Ecological Factors

Climate—Maize has a very remarkable adaptability to nearly all environmental conditions. It is, therefore, a widely distributed crop in the world

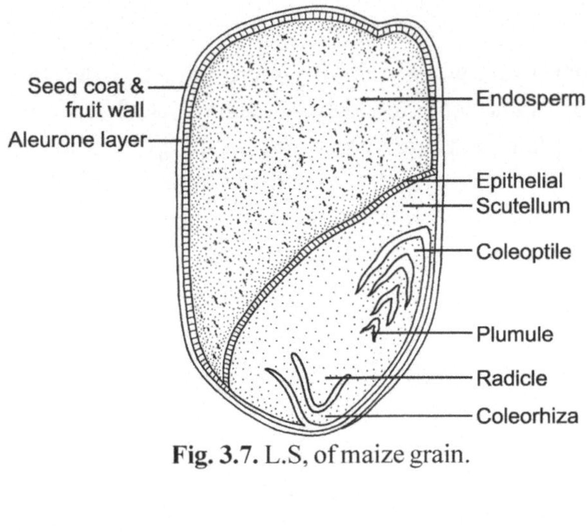

Fig. 3.7. L.S, of maize grain.

extending from the latitude of 58° North in Canada and Russia to 40°S in the Southern Hemisphere. Rainfall varies from 10" in the semi-arid plains of Russia to 200" in India. Many varieties of maize do not germinate satisfactorily at temperatures below 50°F. It is a warm weather crop requiring high temperature during the growing season.

Soil—High yields of corn may be obtained on fertile and well drained loam soils. Soils of Indo-Gangetic regions are well suited for maize cultivation. It comes up well in **Praire soils** and **Chernozems**. The crop is highly susceptible to water-logged condition which results in deficient soil aeration. It is grown well over a range of soil reactions from pH 5 to 8. Good yield can be obtained on alkaline soils which are under irrigation.

Cultivation

The field is cleared of weeds. It is ploughed four to six times. Each ploughing is followed by planking. The soil is pulverized for preparing the seed beds. The soil should, however, remain fairly compact. Seed beds are usually 60 to 75 cm apart. Maize requires heavy manuring. Natural or synthetic fertilizers are given according to the needs of the soil.

Maize is grown as a **kharif crop** in our country. It needs plenty of water during the early stages of its growth. It can be grown unirrigated in those areas which have a rainfall of 75 cm. or more. It needs artificial irrigation when grown in dry summer. The first watering should be delayed as long as possible.

It is usually sown in the rainy season, i.e., in the months of June and July. In the plains of north Bihar bordering Nepal maize is sown much earlier and the crop is ready for havesting in the month of June. Similar is the case in the hills of the North.

Harvesting

The crop is harvested when it is fully matured, i.e., when the sheaths turn brownish and the grains become fairly hard and dry. The cobs are removed from the standing plants and are spread out on the ground in the open sun so that the grains become perfectly dry. The dried up cobs are beaten with sticks to separate the grains. The green dry plants are used as fodder. Quite commonly unripe ears are removed, roasted and consumed.

Kinds of Maize

Several varieties (usually seven) of maize have been recognised (Fig. 3.8).

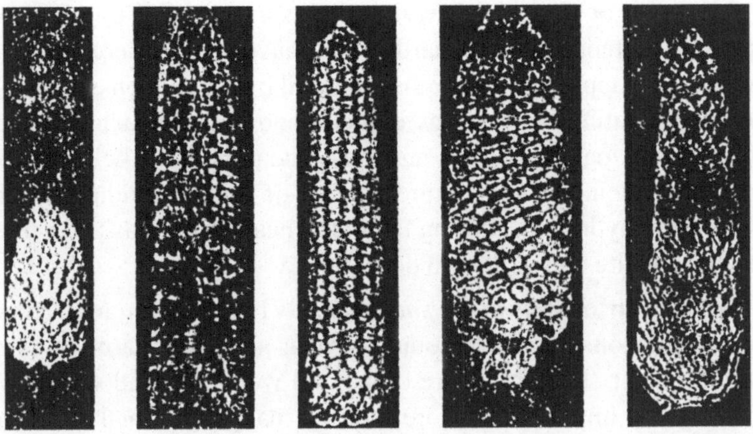

(a) (b) (c) (d) (e)

Fig. 3.8. Kinds of maize: (a) pop corn, with pointed grains (b) flour corn, with variously coloured grains (c) flint corn (d) an ear showing some shrunken grains with sugary endosperm (as found in sweet corn) the others being of the dent type and (e) a modern "pop corn" (By Courtesy of H.G. baker, 1964).

Though these varieties vary in height, growth period and grains characters, the basis of classification is mainly the character of the endosperm and the presence or absence of husk around the individual grains (Fig. 3.9).

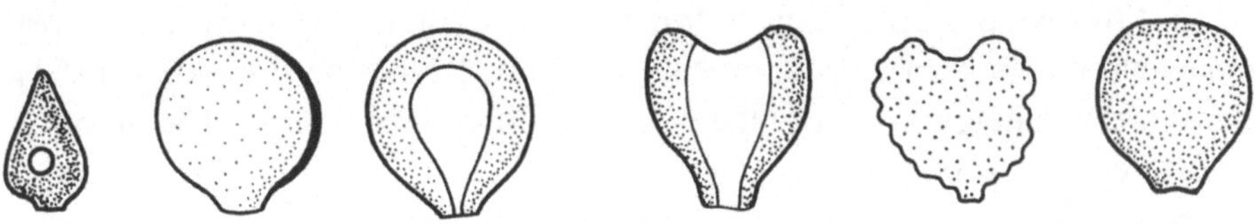

(a) (b) (c) (d) (e) (f)

Fig. 3.9. Diagrammatic sketches of sections through maize grins: (a) pop corn with hard starch (b) flour corn with soft starch (c) flint corn with hard and soft starch (d) dent corn with hard and soft starch (e) sweet corn with sugary starch and (f) waxy corn with waxy starch (By Coutesy of H.G. Baker, 1964).

1. Pod corn or ***Zea mays***—It is supposed to be a primitive form from which the other types of maize have evolved. Each grain is enclosed in a pod or husk while the whole ear is also covered by husk. This species is commercially useless because of the presence of individual husks.

2. Soft corn or ***Zea mays amylacea***—It is called soft corn since it does not possess hard endosperm. The kernel consists of almost entirely of soft starch with a very thin layer of hard starch on the sides. Grains of all colours are found but white and blue are the most common. Soft corn is grown on a very small scale.

3. Pop corn or ***Zea mays everta***—It has exceptional popping qualities. The small elongated, oval and hard grains have a high percentage of hard endosperm. The popping of the grains is caused by the sudden expansion of the endosperm on heating. The grains explode and are turned inside out and a snow-white fluffy, palatable mass emerges. The popping is believed to be due to a rapid expansion of the moisture content of each individual starch grain, after their partial hydrolysis, during the initial heating period. The kernels vary in colour from white and yellow to deep purple.

4. Sweet corn or ***Zea mays saccharata***—It is the common maize. The grains possess a considerable amount of sugar since the starch is completely changed into sugar. The grains are broad and wedge shaped with a wrinkled surface. It is a very important food crop and is usually consumed in the immature (green) stage. Sweet corn is either white or yellow.

5. Dent corn or ***Zea mays indentata***—Dent corn is so-called because of the dent or pit like formation on the top of the grain caused by the shrinkage during the ripening of the inner soft endo- sperm which extends upto the apex of the grain. The hard endosperm is present only on the sides. The indentation varies from smooth to deep or rough type. Dent maize is the principal kind of maize grown in the U.S.A. Kernel colour varies from white to yellow. The dent corn plant grows to a height of 8 to 15 feet and produces only one ear.

6. Flint corn or ***Zea mays indurata***—The grain is very hard. The embryo and white endosperm are completely enclosed by the hard endosperm. The plants are of a height of 5 to 9 feet and are normally two eared. The grains are of different colours.

7. Waxy corn—It is a rare variety of maize. The endosperm is waxy in texture and consists entirely of amylopectin whereas ordinary corn starch is a mixture of 71% to 72% amylopectin and 28% to 29% amylase. Waxy starch takes red stain when treated with iodine. It can be milled to give a product which is a good substitute for tapioca.

Hybrid Corn

Although we are not sure about the ancestry of maize, its recent history has been carefully studied. Maize has proved to be very successful as far as breeding is concerned. It is with maize that greatest achievements in breeding have been attained during the last quarter of the century. During this period, the development of '**hybrid corn**' has revolutionized agriculture in the United States of America and this revolution is now spreading to the rest of the world. Truely speaking, all maize is hybrid in the sense that it is normally cross-pollinated, but 'hybrid corn' is a special kind of carefully produced hybrid.

Normally the plants are **cross-pollinated** but in many cases it is self-pollinated. When self-pollination takes place continuously for several generations, it results in the loss of vigour and show '**inbreeding depression**'. Inbreeding depression is followed by one of stabilization at a low level of vigour. However, when plants from two such inbred lines are crossed, the resulting hybrid is likely to exceed in vigour, compared to even the original ancestors of the inbred lines (Fig. 3.10).

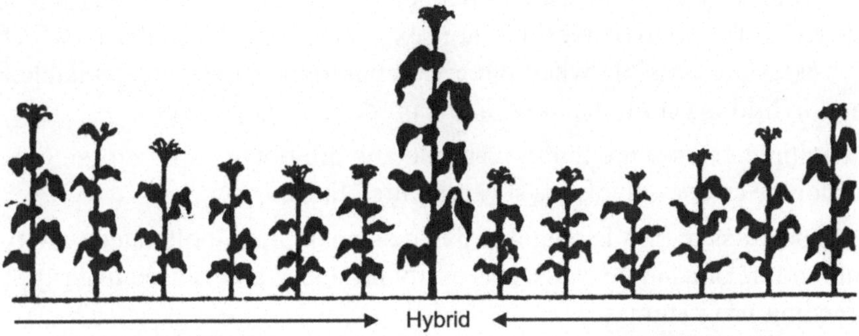

Fig. 3.10. Purification of two lines of maize by inbreeding followed by crossing of the two pure lines to produce an exceedingly vigorous hybrid ("single cross method").

The production of hybrid vigour was first reported by **Charles Darwin** in the middle of the 19th century. **Beal,** a student of **Asha Gray,** carried out experiments designed to used the hybrid vigour for increasing the production of maize. He made a cross between two varieties. When these crossbred seeds were harvested and planted in the next season, the resultant hybrid plants grew vigorously. He did not take plants of pure lines so his results were not fully repeatable and, for this reason, his system was generally not adopted by commercial producers of maize seeds. Several kinds of hybrids have been obtained in maize. They depend upon the number and arrangement of the parent inbred lines and are either single, three way, double, multiple, top or back cross. **George H. Shull** (1908—1909), however, took pure parents (pure lines). Purity in the parents was obtained by inbreeding the parental lines for several generations. The cross between the pure parents gave uniformity in the hybrid product. Only the first hybrid generation was used for commercial corn production. The hybrid obtained by 'single cross' was too expensive for ordinary purposes because it is a poor producer of seeds.

Donald James introduced 'double cross' method in 1918, in which four hybrid inbred strains are combined. This method involves two crossings, which require two seasons before the ultimate seed is produced. 'Double cross' is certainly better than the 'single cross' because single crossed seed is produced from small ears on stunted, inbred plants. Double crossed seed is produced by large ears on vigorous 'single cross' plants. There are not only more ears per plant but also more seeds per year. However, the yield of best single-cross hybrids is much more than the best double crosses.

Double crosses are the most popular of the hybrids and are the most economically produced ones.

A three way cross hybrid is obtained by combining an inbred line and a single cross. This kind of hybrid is more expensive than the double cross. Three-way crosses are possible when three well combining lines are available. This kind of hybrid is very much used in the production of pop corn.

Multiple crosses are highly variable and are obtained by crossing two or more double crosses involving several inbred lines.

A top cross hybrid is obtained by crossing an open-pollinated variety and an inbred line or a single cross. It is easily produced and is used in early stages of breeding programmes.

'Synthetic variety' is the advanced generation of a multiple hybrid obtained by open-pollination. They have several desirable gene combinations which give rise to superior inbred lines.

The types of hybrids with their pedigree is given in Table 3.3.

TABLE 3.3. Kinds of Hybrids of Maize

Kinds of hybrids	Pedigree
Single cross	A × B
Three-way cross	(A × B) × C
Double cross	(A × B) × (C × D)
Multiple cross	(A × B) × (C × D) × (E × F) × (G × H)
Top cross	A × open-pollinated variety or
	(A×B) × open pollinated variety
Back cross	((A × B) × A) × ((C × D) × C)
Single back cross	(A × B × (C × D) × C)
Synthetic	Composite of many lines.

Besides vastly increasing the yield per acre hybrid maize now incorporates desirable qualities of disease resistance, climatic adaptation, and other good characteristics. There is now a large highly organized industry supplying hybrid seeds. These seeds are well adapted to the climatic, soil and disease conditions prevailing in the area in which the crop is to be grown,

A brief account of some of the high-yielding varieties of maize is given below:-

Variety	Description	Area of adaptability
Ganga Hybrid Makka-5	A medium maturing (95—105 days), medium statured, widely adapted hybrid with yellow semi-flint grains. Resistant to brown striple downy mildew (*Sclerophthora rayssae var. zear*) and other foliar diseases, tolerant to stem borer (Chilo zonelus) and drought. Has a yield potential of 45–60 qunitals/ha.	Indo-Gangetic plains and Penisnsular India
Jawhar composite	A high-yielding composite maturing in 100—105 days, with yellow semi-flint grains, resistant to downy mildew and leaf blight; highly resistant to stem borer. It has a yield potential of 40–45 qunitals/ha.	Northern plains and Penisular India
Kisan composite	Medium maturing composite (100—110 days) with yellow semi-flint grains; resistant to downy mildew and leaf blight, has well-developed husk cover; has a yield potentials of 40—45 quintals/ha.	Northern plains particularly Terai belt and eastern Himalayan region.
Vikram composite	Early maturing composite (90—95 days) with orange shining flint grains: tolerant to foliar disease; has a yield potential of 40–45 qunitals/ha.	Northern plains with short growing season.
Sona composite	A medium maturing composite (95—100 days) with orange shiny flint grain; has considerable resistance to foliar diseases and has a yield potential of 40—45 qunitals/ha.	Northern plains.
Opaque-2 Shakti and Rattan composites	Opaque-2 composites with light yellow opaque semi-flint, kernels with 3.4 per cent lysine and 0.95 per cent tryptophan. Suitable for nutritional programme particularly for young children as a substitute for milk. Yield potential of 40—45 quintals/ha.	Shakti-Rajasthan, Terai belt of U.P. and M.P. Rattan—Punjab and Rajasthan.

Punjab Agricultural University, Ludhiana has evolved a new high yielding maize variety called **Ageti-76.** It is easily maturing and is resistant to brown, stripe and downy mildew and is very tolerant to stem borer. It gives 41 quintals per hectare which is 7 pur cent more than Makki Safed-1, released in 1973. Ageti-76 has bold, orange-yellow grains which is preferred to white ones for food. The plant matures in only 88 days.

Uses of Maize

1. New world maize is not very good for making bread since the flour does not have the binding gluten (present in wheat) because of which a cohesive loaf is not baked. Maize is, however, a very useful cereal. In India maize flour is used for making breads in the states like Bihar and U.P.
2. The grains of the ear are roasted and eaten by a large number of people.
3. The green as well as the dry plants are used as fodder. The plant are a very suitable stock feed.
4. The pop-corn and the corn-flakes are well known preparation of the maize grains.
5. Corn starch, corn sugar, corn oil etc. are some of the useful products of the maize grain.
6. The grains are used for making certain alcoholic beverages.
7. Maize is usually fed to the animals to be ultimately consumed as meat or dairy products.
8. The fibres of the maize plant are used for making paper and yarn. Zein, the protein occurring in maize, is now being used for preparing artificial fibres.

Diseases of Maize Plant

The maize crop is subject to the attack of several fungi. The diseases caused by fungi are as follows:

1. Head smut is caused by ***Ustilago reiliana***.
2. Common smut is caused by ***Ustilago maydis*** (***U. zeae***). It causes tumours in place of grains.
3. Downy mildew is caused by ***Sclerospora maydis***. It attacks the young seedlings. Long white streaks appear on the infected parts.

Maize as Rabi Crop

Maize, traditionally a kharif crop can be cultivated to advantage in the rabi season. In fact, the maize kernels produced in rabi are plumpy and comparatively larger in size. Starch and dry milling industry offer a premium price for these corn-cobs since the recovery of starch from them is rather high.

The growers in many districts of north Bihar, eastern Uttar Pradesh, Karnataka and Andhra Pradesh have taken to the rabi cultivation of maize. Farm experts who had mooted the idea based on the results of their experiments believe that the raising of rabi maize can be successfully extended to the entire north Bihar, eastern Uttar Pradesh, Orissa, West Bengal, Western Madhya Pradesh, Gujarat, parts of Rajasthan, Tamil Nadu and Maharashtra– practically in all parts of the country where frost in not a problem in winter. The only hitch is that it can be grown only on fields with assured irrigation facilities as the rainfall in the rabi season is generally inadequate to meet the demand of the crop plants.

Hybrid Strains

"Some of the hybrid varieties which have already been released for general cultivation in rabi include Ganga, Deccan and Deccan-131. Their yield, in some cases, is over 80 per cent more than that of the local varieties. Efforts are continuing to identify more varieties. Many strains now being evaluated possess potential yield about 10 per cent more than the available best composite maize variety, Vijay. The best entry in the experimental material has reportedly yielded about 97.46 quintals of grains per hectare. This is about 41 per cent higher than the yield of Vijay. This material will be used primarily as the parent stock fir breeding new commercial varieties.

The problem of diseases and pests, is not as severe in rabi as in kharif.

Experiments during the past three years have thrown up the possibility of inter-cropping of maize with short-duration varieties of oilseeds and pulse crops. With little effect on the yield of the main crop, the inter-crop proves a bonus. Indications are that it may be possible to grow wheat in between rows of maize. More data is, however, required to substantiate this possibility.

RICE

(Vern. Chaval)

(Oryza sativa)

Origin and History

Rice is one of the most important crops of the world. It is the principal food of half of the world's human population inhabiting the humid tropics and the subtropics.

Rice grain is highly nutritive. It is rich in starch but has a low content of protein (7%). It also contains fat, minerals and vitamins. The aleurone layer is rich in nutrients like protein, fat, calcium and phosphorus. The embryo or the **germ** is rich in protein, fat, vitamins and minerals. The rest of the grain is rich only in starch. The rice endosperm is highly digestible.

De Candolle (1886) because of the presence of a large number of wild forms of rice in **India** thought it to be the original home of the rice. According to **Vavilov** (1930) rice originated in **southwestern part of the Himalayas**. The archeological records, however, show the time of origin of rice in India to be around 2500 B.C. whereas **Chinese** excavations show rice cultivation there in the beginning of the third millenium B.C. On the other hand archeological findings in **Thailand** of rice glumes date back to 3500 B.C. and, therefore, some workers consider rice to have originated in **south-east Asia**. The occurrence of the two species of cultivated rice, the Asian (***O. sativa)*** and the African (***O. glaberrima)*** has led some workers to propose an **African origin of rice**. Asian rice, ***O. sativa*** is grown much more widely as compared to the African species.

Most of the workers including **Chang** of **International Rice Research Institute** (IRRI) believe the two species to have arisen from a common wild species ***O. perennis*** which is still worldwide in distribution. **Chang** (1976) believes this common progenitor to have existed in the humid zone of **Gondwanaland continents** before they broke up and drifted apart. This view is substantiated by the fact that the wild relatives of the two species exist in Africa, south and south-east Asia, Oceania and Australia, and Central and South America. One such species, ***O. rufipogon***, a perennial relative of ***O. sativa*** is found in deep water swamps of southern and south-east Asia, southern China, Oceania and South America. Another wild relative of ***O. sativa***, the annual ***O. nivara*** is very common in southern India, many parts of south-east Asia and Oceania. Several hybrids of ***O. sativa*** and its two wild relatives represented by ***O. nivara*** occur throughout southern and south-east Asia in rice fields and in canals and ponds which are close to the rice fields. ***O. glaberrima*** is also believed to have arisen from an annual weed, probably ***O. barthii***.

On the basis of the wide distribution of wild perennial ***O. rufipogon*** and annual ***O. nivara*** the cultivated rice is believed to have been domesticated first in northeastern India, northern Bangaladesh, Burma, Thailand, Vietnam and Southern China. The three races of rice *viz.,* the Indicas,Japonicas, and the Javanicas spread from this primary centre of domestication (Fig. 3.11). The tropical Indicas spread to Yangtze River Valley in about 200 A.D. The temperae Japonica evolved in the Yellow River Valley of China after rice was introduced there. This rice later spreed to Korea and Japan (300 B.C.) from China. The tropical rice is believed to have later given rise to the Javanica of Indonesia from where it spread to Philippines, Taiwan and Japan. It was only a few centuries ago that rice was introduced to Europe, South and Central America. ***O. glaberrima*** is today limited to only tropical West Africa.

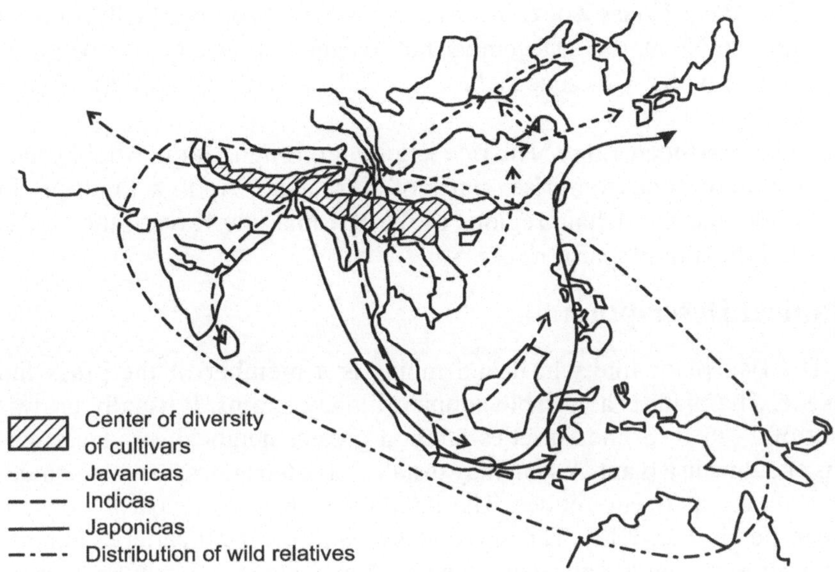

Fig. 3.11. Evolution and spread of geographical races of ***Oryza sativa***.

Rice Growing Regions

World—About 95% of world's crop of rice is produced and consumed in the Orient—in countries like India, Bangla Desh, China, Japan, Korea, Philippines, Malaysia, Indonesia, Thailand, North and South Vietnam, Cambodia, Sri Lanka, etc. Other rice-producing countries are Brazil, U.S.A., Egypt, West Africa, Spain and Italy.

India–It is one of the principal food crops of India. It occupies a vast cultivated area **(44 MH)**. It is mainly confined to the wetter parts of the country. It is chiefly grown in Assam, West Bengal, Bihar, Uttar Pradesh, Orissa, Madhya Pradesh, Punjab, Maharashtra, Karnataka, Tamil Nadu and Kerala. The production of rice is about **92 million tons**.

There are **five** rice growing region in India on the basis of **rainfall, growing season**, etc.

1. **The North Eastern Zone.** The season of the rice crop is from June-July to November- December. It includes Assam, West Bengal, Orissa and southern districts of Bihar.
2. **The Northern Zone.** The rice crop is grown during the period from May-June to September-October. This zone comprises Jammu-Kashmir, Himachal Pradesh, Punjab, Uttar Pradesh and North Bihar.
3. **The Central Zone.** The rice season is from June-July to November-December. The central zone includes Madhya Pradesh, Maharashtra, eastern Karnataka and Telangana region of Andhra Pradesh.

4. **The West Coast Zone.** The rice growing season is usually from May-June to November- December but irrigated crops can also be grown in other months. This zone includes coastal Gujarat, Maharashtra, Karnataka and Kerala.
5. **The Southern Zone.** The rice is grown throughout the whole year. The southern zone comprises parts of Karnataka. Andhra Pradesh, Tamil Nadu and the deltaic regions of the Krishna, the Kaveri, the Godawari and the Tambraparni rivers.

Botanical Description

The rice plant like wheat and maize is a member of the grass family **Poaceae**. The plant is a variable annual of many strains. It usually grows to a height of 2—6 ft. Some varieties attain a greater height. It produces several stems or culms in a plant. A stem may usually have ten nodes but may sometimes have as many as twenty nodes. The internodes are fully or partly covered by the leaf sheath (Fig. 3.12). The internodes are hollow in all varieties but may be solid in shorter plants. The stems of some deep water rice plants grow with the rise of the water level. There are usually seven to twelve tillers developed from the lower nodes of a good strain of rice plant.

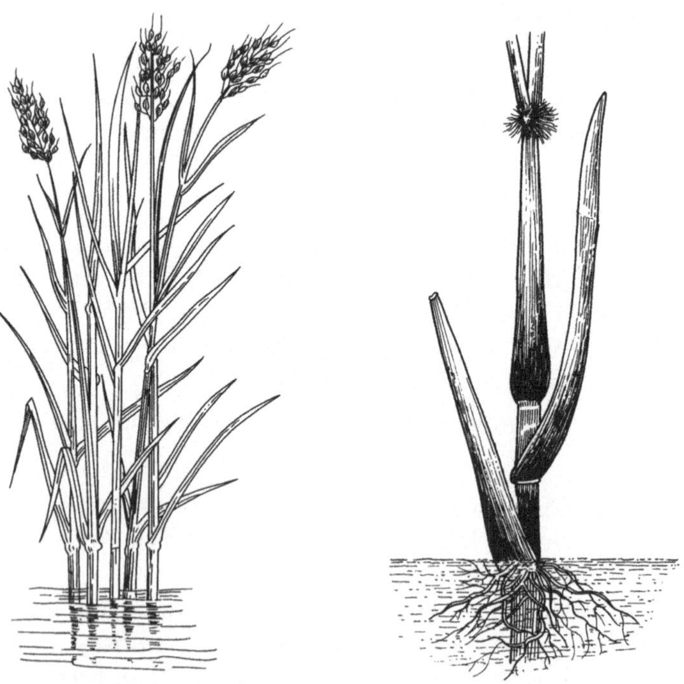

A B

Fig. 3.12. ***Oryza sativa***. A. Plant showing natural habitat; B. Basal portion of the plant.

The inflorescence is a panicle (Fig. 3.13). The spikelets are one flowered and are laterally compressed. It is a self-pollinated flower. The lemma and palea together are known as 'hull' and the rice grain with the hull is known as paddy. The flower contains six functioning stamens which distinguish it from the other cereals. The lemma and the palea may or may not possess awns. The presence or absence of an awn is a hereditary character. The length of the awn is, however, affected by environmental conditions. The presence of awn in rice like that of wheat is considered to be an advantage since it is less subject to bird attacks.

The rice grain is a one seeded **caryopsis**. The rice grain is of various textures, colours, sizes and shapes. The colour of the grain helps in determining the quality and variety of rice. The grain with red colour is supposed to be inferior to others.

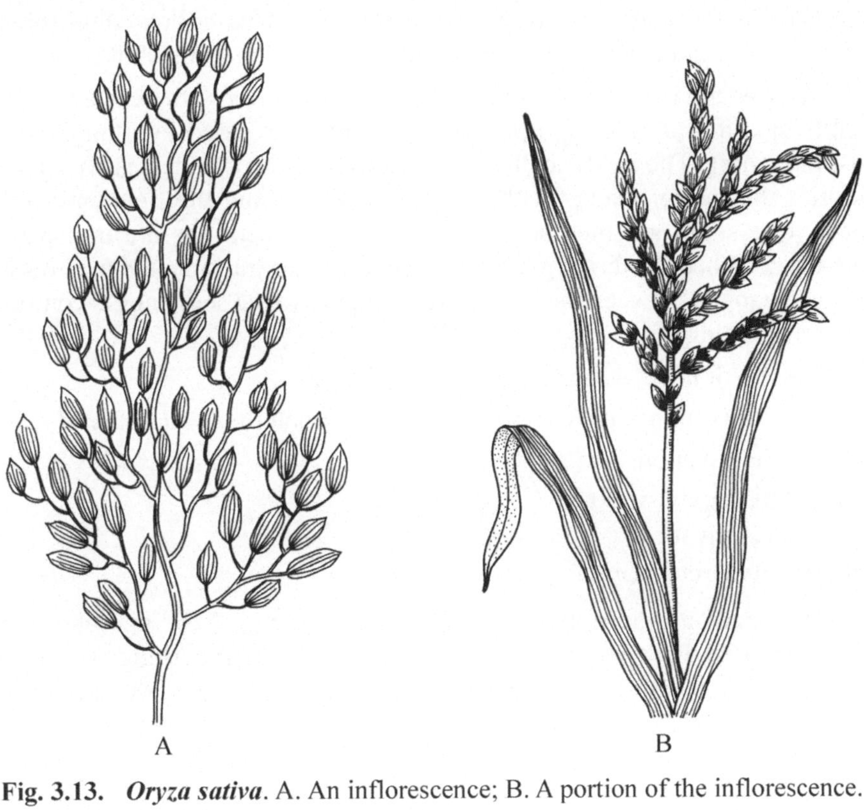

Fig. 3.13. ***Oryza sativa***. A. An inflorescence; B. A portion of the inflorescence.

Ecological Factors

Climate–The plant is grown in plenty of moisture conditions. It is considered to be an aquatic or semi-aquatic plant because of its very large water requirements. **Upland** rice can be grown in areas where annual rainfall is around 48". **Lowland rice** plants are kept submerged throughout its growing period. This prevents weed growth. Submergence enhances **fixation of nitrogen** by blue green algae.

Rice requires warm temperature throughout the growing season with average summer temperature of 77° F.

Soils–Heavy soils-clay or **clayey loam** are well suited to the rice crop because they have very good water holding capacity. Rice can, however, be grown on almost all kinds of soils, *e.g.*, alluvial, red, black, saline, alkaline, peaty and marshy, etc. The soil pH-may range between 4 and 8.

Cultivation

Rice is grown in either standing water (lowland rice) or on dry land (upland rice). The lowland rice accounts for 90 per cent of the world cultivation of rice. Rice is capable of growing in water on account of its hollow stem through which oxygen can easily pass in a downward direction to reach the roots.

The lowland rice is cultivated under flood conditions whereas upland rice is cultivated during monsoon period. The land is prepared by ploughing and cross ploughing. The fields are then flooded with water or if there is no irrigation facilities the farmer waits till the monsoon sets in and the fields are full of water. Rice seeds are then sown broadcast. The seedlings are transplanted from the seed beds to the larger fields. Seedlings are transplanted in rows with adequate gaps in between them. The water supply to the seedlings is controlled after transplating.

Water is drained at the critical tillering stage.

Rice cultivation is of four types depending upon the availability of water.

(a) Dry cultivation
(b) Semi-dry cultivation
(c) Wet cultivation
(d) Intensive cultivation.

The weeds are uprooted by hand. "Green manure" or superphosphates are usually used. Rice crop normally does not bear heavy manuring but breeding for better response to heavy manuring has resulted in the production of high yielding varieties. ***Oryza sativa*** var. ***Indica*** (disease resistant) was crossed with ***Oryza sativa*** var. ***Japonica*** (high yielding and very responsive to fertilizers)." The hybrid showed the desirable characters of both the varieties and the yield increased by 50 to 100%. A rice plant completes its life cycle within 80 – 200 days.

India has adopted Japanese method of cultivation of rice. This method is nothing but a synthesis of improved cultural methods, sufficient manuring and proper protection measures.

Harvesting and Milling

When the ear is ripe and the plant turns yellow the crop is harvested. The seed heads are cut with sickle. The spike is threshed in which the grain is separated from the seed head and chaff and the seed is freed from surrounding hulls or bran. The **bran** is a powdery mass consisting of aleurone layer (protein, 5–10%) and the outer layers of the endosperm (carbohydrate, 40–50%). The process may be done by machines or by hand pounding. The grain is separated from the chaff by winnowing. The rice grains obtained in this manner are brown coloured. The brown covering is rich in nutritious proteins and minerals. As the coloured rice is not good-looking the grains are polished in rice mills. Such polished rice has only the white starchy endosperm left for human consumption whereas the protein rich outer layer finds its use as the cattle feed.

Breeding Rice

Breeders have been engaged in breeding new varieties of rice for improving grain quality, yield and resistance to diseases and pests since the beginning of the present century. The real break-through in the breeding programme was achieved in 1950s when a semi-dwarf **Taichung Native 1** was obtained in Taiwan. This variety had short straw and extensive tilling. India adopted the variety during 1965–66.

International Rice Research Institute was established in 1962 in Los Banos, Philippines with the aim of breeding high yielding variety of tropical rice. The famous semi-dwarf **IR-8** was obtained as a result of cross between the tall tropical variety Peta (Indonesian) and the subtropical semi-dwarf **Dee-geo-woo-gen** (Taiwanese). The variety has several desirable characteristics. It is 100 cm tall, erect, profusely tillering and early maturing plant with stiff culms, dark green leaves, high ratio of grain to straw. It has a high response to nitrogen and no response to photoperiod. The semi-dwarfs **IR-8**, **IR-20**, **IR-22** and **IR-36** are today the most widely cultivated varieties of rice in the world, particularly in the tropics where water control is effective and climatic conditions are favourable.

Oryza sativa var. ***Indica*** and ***Oryza sativa*** *var.* ***Japonica*** are the two major types of cultivated rice. The former is grown in tropical zones and latter in subtropics

High yielding Varieties of Rice

IR–8 (1968)
Jaya–(IET–723) (1968)
Padma (GR–28–25) (1968)
Pankaj (IR–5–114–3–1) (1969) Jagannath (BBS–873) (1969)
IR–20 (1969)
Bala (CR–42–38–173) (1970)

Kanchi (IET–400) (1970)
cauvery (IET–355) (1970)
Krishna (CR–1–6) (1970)
Vijaya (CR–10–5437) (1970)
Ratna (CR–44–11) (1970)
IR–28, IR–29, IR-30, IR-36
Pusa–33 (1975).

PUSA 4-1-11 (IET 1722)

This is probably the first fine grain, non-scented high yielding variety among the very early maturing varieties released so far. The variety which matures in about 103 days gives an yield of 5500 kg per hectare. It does well under late planted (aged nursery) condition. It is moderately resistant to major diseases. Long slender grain with high milling and low breakage percentage and excellent cooking quality are its other characteristics. It gets relatively higher price than the coarse varieties in the grain market. By virtue of its short duration it may also serve as a good choice for various multiple cropping sequences.

A brief account of some of the high yielding varieties of paddy is given below:

Variety	Description	Area of adaptability
Jamuna	A dwarf variety, having long slender white grains with good cooking quality. Cooked rice is non sticky. Maturing in 110 days and gives an yield of 40–50 quantals/ha.	Punjab, Haryana, Delhi, Western U.P., M.P. and Rajasthan (canal commanded areas).
Sabarmati	A dwarf variety having medium slender grains with fine scent and good cooking quality. Cooked rice non sticky. Matures in 120 days and gives an yield of 50–55 quintals/ha.	Delhi, Punjab, Haryana, and U.P., canal commanded areas of Rajasthan and summer sowings for Bihar.
Improved Sabarmati	Improved selection of Sabarmati. Grain finer than Sabarmati and have fine aroma. Good cooking quality. Matures in 130–135 days and gives an yield of 60–65 quintals/ha. Suitable for export.	Delhi, Punjab, Haryana and U.P., and also canal commanded areas of Rajasthan and areas where there is demand for scented rice,
Pusa-2-2] **(Kannagi)**	Dwarf variety having short bold grains. Good cooking quality. Suitable for transplanting as well as direct sowing. Matures in 100 days and is suitable for multiple cropping yield, 55–60 quintals/ha.	Tamil Nadu (for 'Kuruvai' season), Kerala, Tripura, Assam, Bihar,) (short season), Delhi, Punjab, Haryana. etc. (for multiple cropping.

(*Contd.*)

Variety	Description	Area of adaptability
Pusa, 33	Shortest duration rice variety. Matures in 100–105 days, suitable for multiple cropping; grains long and slender, translucent and scented. It combines the fine aroma and looking qualities of basmati with the high yielding characteristics of the dwarf types. Its yield is almost three times of true basmati varieties. Average grain yield 52 quintals /ha.	Haryana, Western U. P., Bihar, Maharashtra, Tamil Nadu, Assam and other North-East India States,

Uses of Rice

1. Rice is only one of the great cereals which is consumed as harvested (after dehulling and perhaps polishing). It is simply boiled for direct consumption. Rice with some legume (pulses etc.) form the diet of millions of Asiatic people,

   Rice grains are of two kinds—**glutenous** which is sticky when cooked because of the presence of dextrins and **nonglutenous** which is hard and 'flinty'. The nonglutenous types are by far the most important varieties.

   Rice grains are classified on the basis of kernel shape into short (about 5.5 mm), medium (about 6.6 mm), and long (7 to 8 mm) types. In India people prefer long grains. Most of the short grain types are grown in Japan and Taiwan. They are sticky when cooked. This nature is preferable in these countries because they eat rice with chopsticks.
2. The parched rice and the beaten rice are normally consumed by the poorer section of the society.
3. Some cooked rice preparations are Dosa, Iddli, Pulao, Kheer, etc.
4. Refined bran oil is used as salad oil.
5. Bran oil is used as carriers of insecticides. It is also used in textile and leather treatment. It is used as lubricant in wrist watches etc. because of its anti-corrosive and rust resistant properties. It is also used in the manufacture of soap.
6. The husk and rice polish are good for stock-feeding.
7. A byproduct of rice cooking called 'Mar' is used by washermen for stiffening of the washed cloth.
8. Rice flavour is used in cosmetic industry.
9. The straw is used in the making of hats, shoes and other articles.

Diseases

1. Brown spot diseases of rice is caused by ***Helminthosporium oryzae***. Brown coloured round or ellipsoidal spots appear on the seeds, leaves, leaf sheaths and nodes etc. It is a seed-born disease.
2. Blast of rice is caused by ***Piricularia oryzae***. The nodes become black and pale spots with brown margins appear on the leaves. It is also a seed-borne disease.
3. Stem rot is caused by ***Sclerotium oryzae***.
4. Yellowing of seedlings and mature plants is caused by ***Fusarium***.

OATS
(Vern. Jai)

(Avena spp.)

Origin and History

After wheat, rice and maize, oats constitute the fourth major crop of the world. Oat is a dominant cereal of temperate and Mediterranean region. Oat is highly nutritious. The grain of oat contains 16% protein and 8% fat. Oat species of the genus ***Avena*** are among the most widely grown cereals of North America and Europe, being better adapted to a wide range of soil types and climatic extremes than are the other cereals.

Oats are possibly of **multiple origin**. The oats have been traced to the year 1000 B.C. in **Central Europe**. The hexaploid ***Avena sterilis***, which is one of the wild oats, is believed to have given rise to the hexaploid ***Avena sativa***. On the other hand, ***A. sativa*** is believed by some to have originated from another hexaploid ***A. fatuva***. Since ***A. sterilis*** is widely distributed around the Mediterranean and extends through the Asia Minor to the slopes of the Zagros mountains in the east in the relatively stable primary habitats and occurs in areas of its tetraploid and diploid progenitors, it is believed to be the real ancestor of ***A. sativa. A. sterilis*** is also believed to have given rise to ***A. byzantina*** (Fig. 3.14). The ***saliva***. and ***byzantina*** are believed to have given rise to the free-threshing form, *A. nuda* which extends eastwards from the centre of origin to China. The cultivated diploid forms of ***strigosa*** group are supposed to have evolved in Iberia in the western Europe. The ***barbata*** group of tetraploids of West Asia is supposed to have originated from the ***strigosa*** diploids. The cultivated tetraploid ***abyssinica*** of Ethiopia is believed to have arisen from the weedy *barbata* types.

While the diploid and tetraploid species are of very little importance the large grained ***sativa-byzantina*** hexaploids are widely grown as the major cultivated cereal. From Europe it spread to North America, Argentina and Australia.

Oat-growing Regions

World—About 80% of the oat-growing regions lie in the northern temperate zone. The chief oat-producing regions are North America, Europe, Russia and Canada.

India—The crop was introduced in India in the recent past. It is chiefly grown in Uttar Pradesh, Punjab, West Bengal, Madhya Pradesh and in some parts of Maharashtra.

Botanical Description

Like other cereals oats also belong to the family **Poaceae**. The oat plants attain a height of 2 to 5 ft. The stem has 4 to 8 nodes with tillers developing from the base during early growth. Secondary and tertiary tillers up to 30–40 may be found in winter varieties. The lower portion of the plant is weaker than the upper portion of the stem and therefore, lodging of stems is quite common. The lower nodes produce adventitious roots. There are many bluish green leaves on an oat plant.

The leaf sheath closely envelops the internodes (Fig. 3.15). The inflorescence may be erect or drooping. It is a much branched panicle, sometimes expanded and sometimes condensed and one sided. The ultimate branches end up in about 75 spikelets which are completely covered with glumes. The spikelets bear one to seven florets. The upper ones are often staminate or sterile. The lemmas may or may not be awned. The florets are self-pollinated. The oat grain is a caryopsis fruit.

Ecological Factors

Climate—Oats are adapted to cool moist regions of the world. They can, however, grow in extremely cold places as well. Red oats are grown primarily in warm climates, *viz.,* southern United States, Mediterranean region of Europe and Africa, Australia and Argentina. Oats require more water than any other cereal crop. Drought often causes great damage. High temperature does not suit oats unless they are given plenty of water. High temperature during flowering lead to the formation of empty spikelets commonly called 'blast'. Hot, dry weather at the time of ripening is deterimental and leads to premature ripening and hot humid weather favours prevalence of certain diseases, which reduces grain yield.

Soil—As pointed out earlier, oats can be grown in a very wide range of soil types. They are less sensitive to acidity and can be grown on a wide range of soils with reasonable drainage and fertility. Oats prefer heavy soils, but they can be grown even on poor and exhausted ones.

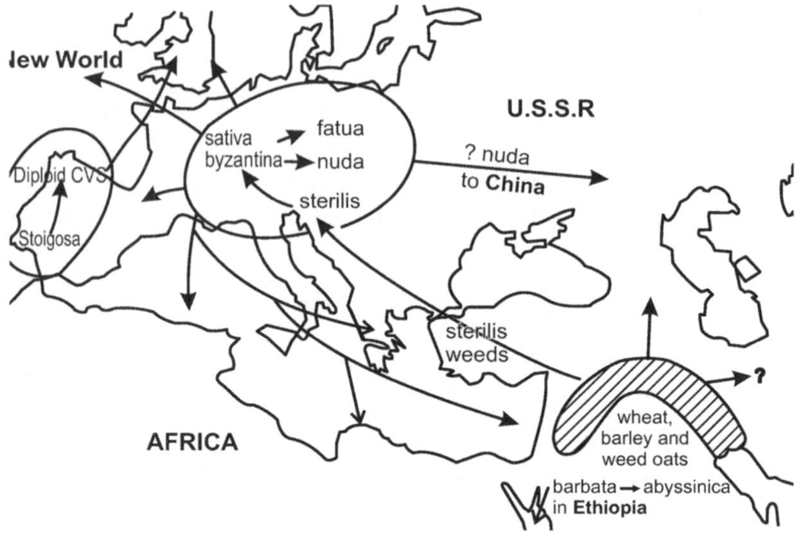

Fig. 3.14. Evolutionary geography of the cultivated oat, *Avena spp.*

Cultivation

Oats like wheat have two varieties–the winter and spring varieties depending upon the season of planting. In India oats are cultivated as a **rabi crop**. The land is prepared in a way similar to that of wheat. It is ploughed three or four times. The seeds are sown soon after the cessation of rains in the month of September or October.

Oats are generally planted with other crops. Oats should preferably follow a crop which is deep rooted one. The yield of oat is good if the previous crop is manured. The fodder crop may be supplied with ammonium sulphate or oil cake.

The sowing season extends from October to December. In the case of grain crop the seed is sown in rows 25 to 30 cms apart whereas in the case of fodder crop seeds are sown by broadcast method.

Harvesting

The crop matures in 3 to 4 months. The plants are harvested when the leaves are still green and the grains are not fully ripe. The harvested crop is left for a few days for final drying and maturing of the grains and is then subjected to the threshing process in a manner similar to that of wheat and barley.

Kinds of Oats

There are three broad groups of oats classified on the basis of the number of chromosomes.

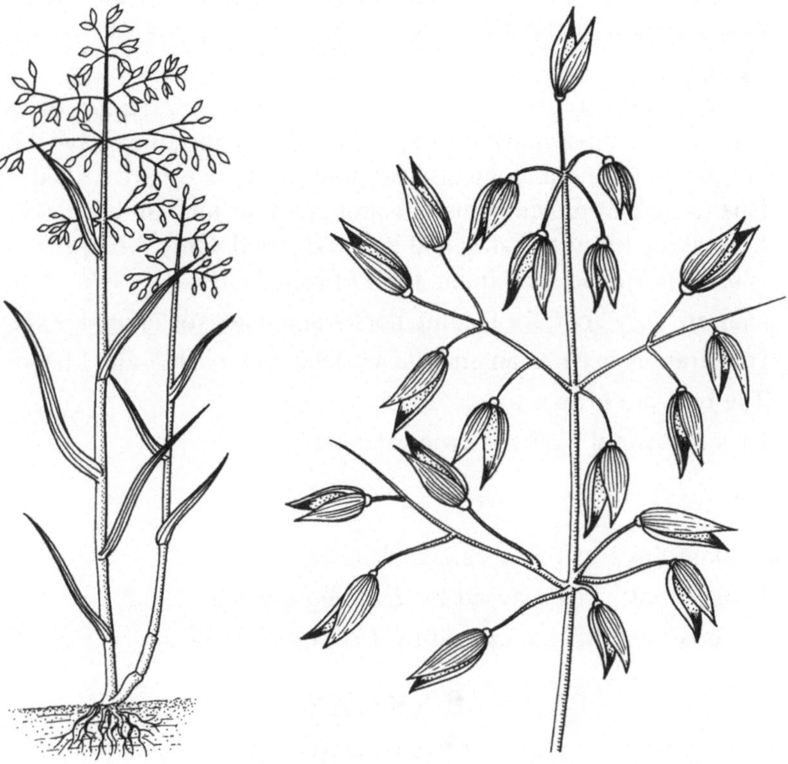

Fig. 3.15. Oat. A. Plant; B. Inflorescence.

A. Hexaploid species (2n = 42).

1. ***A. sativa***—most important cultivated oat.
2. ***A. fatua***—troublesome weed in oat fields.
3. ***A. sterilis***–similar to ***A. fatua*** but with larger spikelets.
4. ***A. byzantina***–warm climatic, cultivated oat.
5. ***A. nuda***—lemma and palea incompletely fused with the grain, yielding a naked seed easily on threshing. Not much cultivated.

B. Tetraploid species (2n=28).

1. ***A. barbata***–tall, weak, slender, ranch grass of the western United States.
2. ***A. abyssinica***—grown for forage in the desert areas of northern Africa.

C. Diploid species (2n = 14).

1. ***A. brevis**-the* short oat.
2. ***A.* strigosa**—the sand oat.
3. ***A. wiestau***—the desert oat, and
4. ***A nudibrevis***.

The diploid oats are of little commercial importance and they are cultivated to a limited extent in parts of Europe.

Uses of Oats

1. Oats are the most nutritious of all cereals because they have a high fat, protein and mineral content. The flour of oat is usually called oatmeal. Due to lack of gluten oatmeal is not good far making breads. It is used for making biscuits, cakes and breakfast food. Only 3 or 4 per cent of it is used as human food in the form of rolled oats.
2. Oats are very good for feeding horses and domestic animals except pigs.
3. Oat straw is used as an emergency feed and as an animal bedding
4. The oats are grown for green forage.
5. Oats are useful plants in crop rotation.

Diseases

The important diseases of oats are as follows:

1. Loose smut of oat caused by ***Ustilago avenae***.
2. Covered smut of oat caused by ***Ustilago kolleri***.

BARLEY
(Vern. Jau)

(Hordeum vulgare)

Origin and History

Barley is one of the most ancient of cultivated plants. It is argued that barley was used for bread before the advent of wheat. It is considered to be the **oldest** of all cultivated plants. There are enough evidences to prove that the crop was known to the **Chinese**, the **Egyptians**, and the lake dwellers of **Switzerland** in the very remote past. This crop must also have been known in our country when the Aryans first came. The Sanskrit name 'Java' (Javaka), which originally meant a grain, and was later used only for barley, and from which the name 'Jav' and 'Jau' has been derived, shows that barley was in those days regarded as one of the most important grains. **Vavilov** was of the opinion that barley might have originated in the arid lands of southwestern Asia, southeastern Asia and northern Africa. It can grow on high mountains, in the Arctic zone and even in desert oases.

The remains of wild barley have been found at Mureybat on the Euphrates in Syria (8000 B.C.), at Beidha in the southern Jordan highlands (6800 B.C.) and Jarmo in the Zagros foothills in Iraq (6800 B.C.). Remains of cultivated barley have been found in several areas of West Asia *viz.,* Ali Kosh (Iran), Jericho (Palestine), Ramad (Syria), and Turkey in between 6000 and 7000 B.C.

Barley is believed to have reached Spain in the fifth millenium B.C. It reached Indus in the third millenium B.C. and China in the later part of the second millenium B.C. Later it spread to Ethiopia from West Asia.

There are archaeological evidences to suggest that as compared to wheat, barley was a much more important food crop of the ancient Egyptians. Barley was introduced into the New World by Columbus. It reached cold highlands of Mexico in the beginning of sixteenth century A.D.. Subsequently it spread to Peru, Chile, North America and Canada where barley is chiefly grown as feed for the livestock. The world production of barley is around 170 MT. per annum.

It is today one of the most important cereals of the world. Barley provides food for large number of people, particularly to the poorer section of society.

Barley Growing Regions

World–It is cultivated in almost all parts of the world. The principal barley producing countries are China, Japan, U.S.S.R., Rumania, the Mediterranean fringe of North Africa, Ethiopia, Turkey, the United States of America, Canada and India. India is one of the most important barley producing countries of the world.

India–In this country the bulk of the crop is produced in the Indo-Gangetic plains. The area of greatest production are Uttar Pradesh, Rajasthan, Bihar and Punjab. The States of Uttar Pradesh and Rajasthan produce approximately 73% of India's total production of barley.

Botanical Description

Barley belongs to the genus ***Hordeum*** and like all cereals is a member of grass family **Poaceae**. As a plant, it very much resembles wheat, but differs slightly from the latter in the length of the stems, the character of leaves, and the structure of the spike (Fig. 3.16A).

The stems possess five to seven internodes. The height of the plants varies with varieties and also with environmental conditions. Generally the average height is from 2 to 3 ft. which is less than that of wheat plant. The degree of tillering is usually much. Lodging is not as common as in wheat. The leaves of barley also resemble those of wheat but are usually broader, and the colour of the leaves is not as deep green as that of the leaves of wheat.

The inflorescence consists of a terminal spike, the shape of which varies slightly in different species called the two-rowed, the four-rowed or six rowed barleys (Fig. 3.16 B and C). In barley three spikelets occur at a node of the rachis, a character which distinguishes it from wheat which bears only one spikelet on one node of the rachis. In the six-rowed type, which is the most common in India, the spikelets are in six distinct rows. All the spikelets are fertile and produce grains. Barley has only one floret in a spikelet, whereas wheat usually contains three or more. In two-rowed barleys the spikes bear two longitudinal rows of grains. This is due to the fact that the outer spikelets at each node are sterile and only the central spikelets produce grains. In this way four of the six rows are eliminated and only two rows are left. Barley is normally self-pollinated and natural cross fertilization seldom occurs. The reason for this probably is that the styles are short and do not protrude beyond the glumes.

The grain of barley is a caryopsis and unlike wheat usually adheres, to the lemma and the palea or hull. In some varieties which are known as 'naked barleys' this condition does not occur.

The root system of barley consists of two more or less well defined types—the shallow and the deep rooted systems.

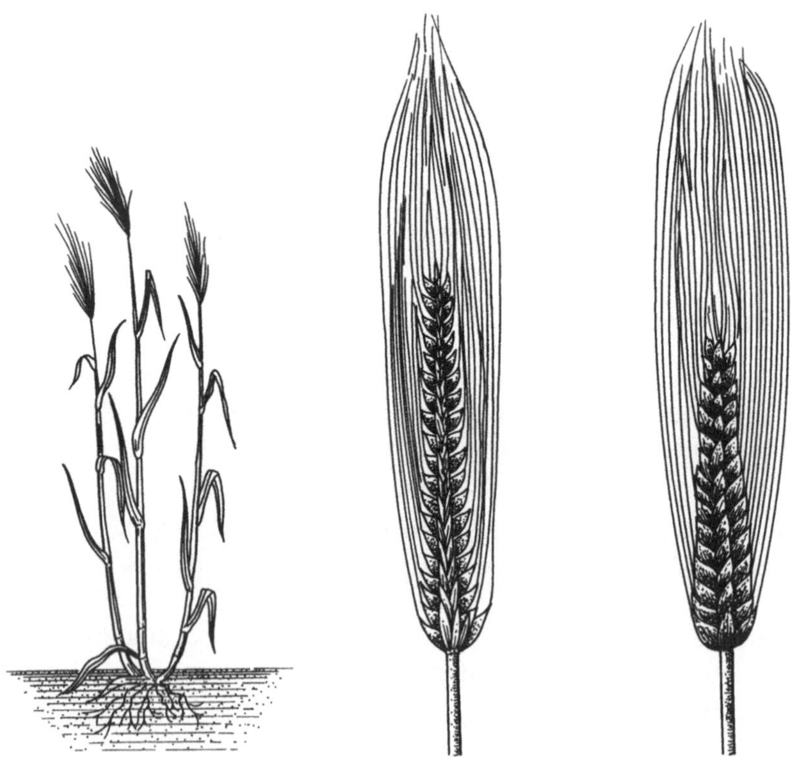

A B C

Fig. 3.16. Barley. A. Plant; B. 2-rowed barley; C. 6-rowed barley.

Ecological Factors

Climate—Barley is a plant that can grow fairly, well in temperate as well subtropical regions of the earth. It is, therefore, adapted to varying conditions of climate. It is essentially a cool-season crop. It does well in regions where winter is cool and the crop growing period lasts for about five months. It needs less water than the wheat or oat crop. It is not suited to warm humid climates. Hot winds induce premature ripening of grain and high humidity may check the proper maturation of the crop. The property of malting is very much affected in both the cases. Discolouration of grain is caused if it rains at the time of harvesting. It is a short season early maturing grain. It gives a high yield. It has adaptability for a variety of climatic conditions.

On the whole it grows well in cool or slightly warm and dry climate.

Soil—Barley is grown on light soils. The well drained, medium loam soil of even texture is the most suitable for its cultivation. It requires more porous soil than wheat. In most parts of Uttar Pradesh it is grown in soils which are too light or otherwise not very favourable for the growth of wheat. Barley can withstand more alkalinity in the soil than wheat.

Cultivation

The preparation of land for the cultivation of barley is not as thorough as in the case of wheat. The number of ploughing is much less in this case. Using an improved plough, one ploughing followed by harrowing is regarded as sufficient for preparing seed-bed for barley. Unlike wheat barley is sown from the middle of October to the middle of November. Late sowings produce grains of poor quality.

Seeds are sown either by broadcasting or in shallow furrows by single tube drill which is attached behind a plough. The rows are spaced about 25 cm apart.

Barley is grown unirrigated in regions having a rainfall of 40–50 cm. It needs two or three irrigations after sowing in drier areas.

Fertilizer application is very effective under dry conditions. Foliar spray of urea is twice as efficient as its soil application. Brewing quality of barley is adversely affected by excess of nitrogen. Weeding is not generally practised.

Harvesting

Harvesting takes place from third week of March to middle of April. The method of harvesting is similar to that of wheat, the plants being uprooted or cut with a sickle when fully ripe. But as the crop shatters more easily than wheat, it is usually harvested in the forenoon before the heads become too dry and fragile for handling. The crop is then taken to the threshing floor where the bullocks are made to tread on the dry produce. The grain is separated from the chaff by winnowing.

Kinds of Barley

There are a great number of present day forms of ***H. vulgare***. Barleys have been classified by several workers since the time of **Linnaeus,** but the classification by **Harland** (1918) seems to be the most satisfactory. It is as follows:

A. Six-rowed: All spikeletes are fertile.

1. ***H. vulgare*** L: Lemmas of all florets are awned or hooded,
2. ***H. intermedium***: Only the central florets have awns. The latteral florets never possess the awns.

B. Two-rowed: Only the central spikelets are fertile.

1. **H. *distichon*** L: Lateral spikelets consisting of outer glumes, lemma, palea, rachilla and usually rudiments of the sexual organs.
2. ***H. deficiens***: Lateral spikelets are reduced to only the outer glumes and rachila, lemma and palea being absent. No rudiments of sexual organs present.

Uses of Barley

1. Barley was the chief source of bread in the 16th century and it has remained a staple food since then. It is highly nutritive. It is now used less for bread making as it contains gluten in small quantity.
2. Barley is used as breakfast food and as children's food. Six-rowed barley is a good source of protein.
3. It is also used as fodder.
4. About 20–30% of the crop is used as a source of malt to be used for preparing alcohol, whisky, beer and similar beverages.

Diseases

Barley is subjected to a number of fungal diseasas of which following are the most important

1. Covered smut caused by ***Ustilago hordei***.
2. Loose smut caused by ***Ustilago nuda***.
3. Powdery mildew caused by ***Erysiphe graminis***.

MILLETS

Millets are small-grained cereals. The tropical millets include the sorghum, pearl millet, finger millet and kodo millet. The tropical millets are capable of drought resistance and can grow even in semi-arid conditions. They have an additional correlated feature *i.e.,* they can grow on poor soils.

The temperate millets include the common millet, little millet, foxtail millet, and Japanese millet.

SORGHUM
(Vern. Jowar)
(Sorghum vulgare or ***S. bicolor)***

Origin and History

Sorghum and bulrush millets are the major cereals of the semi-arid tropics. They are cultivated in rain-fed regions. They constitute the staple food of more than 400 million people. While Sorghum is a food crop in India and Africa it is grown for stock feed in Japan, Europe and countries of the New World.

The genus ***Sorghum*** includes a number of widely cultivated grasses. The most important species of sorghums is ***vulgare,*** which has great use as food. Sorghums have been cultivated by man in the remote past, when he started the domestication of wild species. Sorghum has been reported to be in cultivation in **Egypt** before 2000 B.C. It was also cultivated in **China** and **India** in ancient days. The sorghums are believed to have originated in **Africa**, as well as in **Asia**, where they still continue to be the principal food of many native peoples. The sorghums are believed to have then spread to the warmer countries of the temperate and the tropics. Sorghum was introduced in the United States about the middle of the nineteenth century.

While most of the investigators believe that cultivated sorghums have originated from the perennial Johnson grass ***(S. halepensis)*** or its annual form the Sudan grass ***(S. vulgare*** var. ***sudanensis)***. **Snowden** in his monograph has expressed the opinion that they have arisen from four ancestral plants, *viz.,*

(*i*) ***S. arundinaceum,***
(*ii*) ***S. verticilliflorum,***
(*iii*) ***S. aethiopicum,*** and
(*iv*) ***S. sudanensis.***

Sorghum Growing Regions

World–Sorghums are grown in Africa, India, China, Manchuria and the United States.

India–In the world India has the largest area under (MH) jowar cultivation followed by the African continent. Next only to rice, jowar has the largest area under cultivation in our country. The principal jowar producing states in India are Maharashtra, Tamil Nadu, Andhra Pradesh, Madhya Pradesh and Uttar Pradesh. The total area covered by jowar is approximately 8.51 million hectares and the total annual production exceeds 8.50 million tons in 2006–2007.

Sorghum is an annual plant. It belongs to the family **Poaceale**. The plant varies in height, with different varieties, from 4 feet to 15 feet or more in some tropical forms. In habit and appearance the plant resembles maize (Fig. 3.17).

The stems or culms of jowar are cylindrical and pointed. The number of nodes is different in different varieties and consequently the number of leaves and the length of internodes also vary with different varieties. The number of nodes is usually ten. The thickness of stems usually varies, being about one inch in thicker ones. Most of the jowars in this country do not tiller, but some introduced fodder varieties, however, are capable of producing three or more tillers from one plant. Usually two or three lower buds develop into tillers. The stems of jowar also vary a great deal with regard to their juice contents. Some are quite juicy and sweet, whereas some are pithy and contain very little juice. Certain introduced fodder varieties are more juicy and also sweeter than the common jowars of India.

Fig. 3.17. Jowar Crop (Photo by N. Singh).

Botanical Description

The leaves are similar to those of maize, though they have almost half the total area of maize leaves. The leafy character is desirable for fodder varieties. Like the leaves of the grasses, a jowar leaf also possesses parallel veins and a mid rib which is generally white in colour. Some varieties, however, possess a dull-green mid rib. It has been found that this character is associated with the juiciness of the stem.

The inflorescence of jowar is a panicle with varying degree of looseness or compactness (Fig. 3.18 A). Most of the grain jowars in India may be classed as **compact**, **loose** or **intermediate** types. The panicles are commonly known as '**heads**'. These vary a great deal in size, shape, form and colour. The peduncle, commonly known as the '**neck**', may be straight, curved downwards or half bent. The panicle is composed of numerous spikelets which usually occur in pairs, one of them is sessile and the other one is pedicellate. The sessile one is broad, thick and fertile, whereas the pedicellate one is narrow, long and staminate and hence does not form seed. However, the spikelets occur in threes near the top of the panicle, in which case one is sessile and fertile and the other two are pedicellate and infertile.

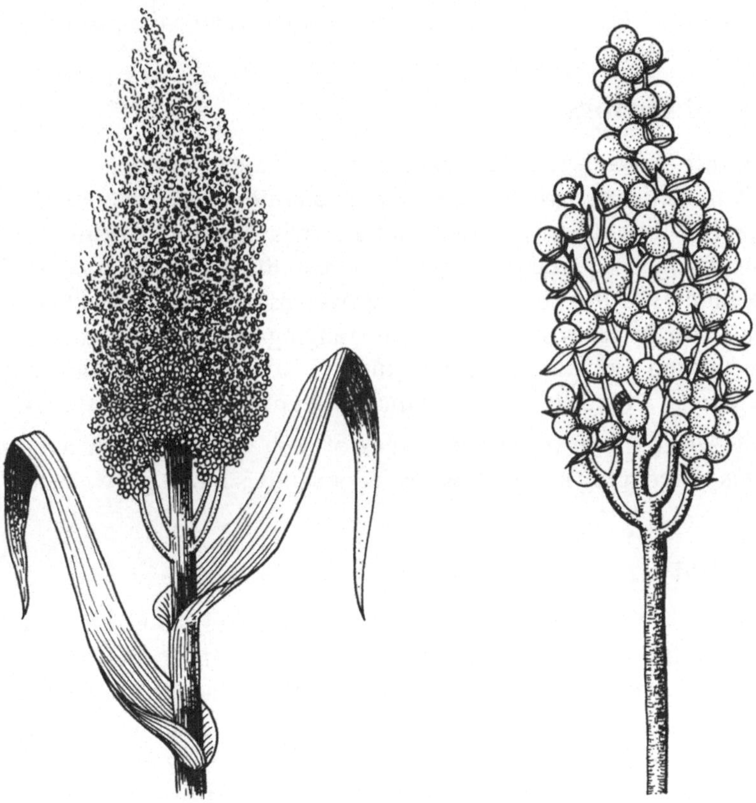

A B

Fig. 3.18. S. ***vulgare***. A. Mature head; B. Magnified portion of the head.

A sessile spikelet has two flowers, one fertile and the other sterile. Generally one grain is formed as a result of the fertilization of the fertile flower. Occasionally, however, more than one grain may develop producing a phenomenon known as multiple seededness. **Ayyangar** reports as many as six seeds in a sigle spikelet, but two is the most common number. In this country the double seeded grain jowar is known as '**do-dania**,. A mature head of ***S. vulgare*** is shown in Fig. 3.18 B.

The jowar flowers begin to open after the inflorescence has come out of the terminal leaf sheath which is known as the **boot**. The first flowers to open are those near the apex and blooming proceeds progressively downwards on the panicle. Normally the flowers are pollinated by the higher flowers of the same panicle.

Jowar possesses a well developed root system similar to that of maize but the roots are generally wiry and fibrous, and also usually, more abundant. The roots are widely and thickly distributed and their absorptive power is so efficient that they are able to exhaust most of the available moisture from the soil. The extensive root system and its great absorptive nature, reduced leaf surface and the ability of the leaves to roll up in dry weather, enable the plant to withstand drought and arid conditions.

Ecological Factors

Climate–Jowar is remarkably drought resistant and prefers warm climates. It is grown in climates which are too dry and hot for maize. In India it grows in those areas which have moderate rainfall. It is grown as a **kharif** crop. The minimum rainfall required from sowing time to the harvesting time is 12 in. and the maximum is 40 in.. The heavy rains as well as prolonged drought damage the crop. The crop thrives best in temperature ranging between 80° and 90" F, though it can withstand a maximum temperature of 105°F to 110°F. The crop can be grown up to an altitude of 3000 feet though it grows best on the plains.

Soil—It is grown on various types of soilranging from heavy and light alluviums to red, grey and yellow loams or sandy soils. It grows better on clay loams.

Cultivation

The preparation of the land for the sowing of jowar is not as thorough as that of cereals. In the black soil area, if the land is badly infested with weeds, ploughing followed by harrowing is usually practised. In the alluvial soil area the land is usually ploughed two or three times before sowing. In Northern India when the monsoon sets in, the seeds are sown either by broadcasting or in rows behind the plough. When sown in lines the distance between the rows varies from 14 in. to 2 feet, depending on the character and fertility of soil. The plants are usually thinned to a spacing of about 9 in. to 12 in. in a row. When jowar is grown exclusively for fodder purposes the seeds are usually sown broadcast.

The first intercultivation is given by hand-hoes about a fortnight after sowing and weeds are removed.

Harvesting

The grain takes four to five months to become ripe. The stalks are then cut as close to the ground as possible by means of sickles. They are then tied in bundles and brought to the threshing floor where the heads are cut-off and the stalks are stacked. When the heads are throughly dry, they are heaped on a threshing floor and the bullocks are made to tread on the grain. A much quicker, labour saving and more efficient method is to use stone-roller or metal disc threshers. The grain in finally separated-from the chaff by winnowing. The stalks are usually cut into small pieces collectively known as **kadbi**, which is used for feeding the cattle.

Storage. To prevent damage from pests etc. the grain is thoroughly dried in sun, cleaned and is then stored in earthenware. A two-inch thick layer of sand is put over top of the grains and the container is plastered with a mixture of earth and cow-dung to prevent ingress of insects. The grain is removed through a trap whenever needed. High yielding varieties are CSH-1, CSH-2, CSH-3, CSH-5, CSH-6, CSH-7R, CSH-8R, SPH-61, etc.

High-yielding varieties of Sorghum

A brief description of some of the high-yielding varieties of sorghum is given below :

Variety	Description	Area of adaptability
CSH-1	A high yielding hybrid of medium height and uniform ripening. Very widely adaptable and suitable for irrigated as well as dry conditions. Matures in 100–110 days and gives an yield ranging from 35–40 quintals/ha.	Tamil Nadu, Andhra Pradesh, Karnataka, Maharashtra, Gujarat, M.P. and Rajasthan.
CSH-2	Slightly taller than CSH-1, matures in 110–120 days, yield range is 30–35 quintals/ha.	Throughout India particularly in the irrigated areas of Tamil Nadu.
CSH-3	Matures in 110–120 days and gives an yield ranging from 30–40 quintals /ha. Grains quality better than CSH-1 and CSH-2.	Maharashtra, Karnataka and A. P.
Swarna	Matures in 100–110 days and gives an yield ranging from 25–30 quintal /ha. Grain quality superior to CSH–1. Being a non-hybrid variety, farmers can keep their own seed.	Throughout India.

Uses of Jowar

1. It is staple food of the poor classes. Grain may be broken and cooked in the same way as rice. Grain sorghums are usually grounded and the flour is used for various preparations. Some varieties are used as popped grains.
2. The stalks of the plants green or dry, provide nutritious fodder for cattle. The stalks are usually cut into pieces before being used as fodder for cattle.
3. It is also used in making some industrial alcohol, whisky, beer and starch.
4. Broom sorghums are used as fibres for making brushes.
5. A wax similar to carnauba is reportedly obtainable from the hulls.
6. Sweet sorghums are used for making syrup.
7. Sweet sorghums are used as forage.

Diseases

1. Red leaf spot is caused by ***Colletotrichum graminicolum***. This is probably one of the most common diseases of jowar in India.
2. Rust is caused by ***Puccinia purpurea***. This is a fairly common disease.
3. Leaf spot is caused by ***Cercospora sorghi***.
4. Downy mildew is caused by ***Sclerospora graminicola***.
5. Grain smut is caused by ***Sphaecelotheca sorghi***.
6. Head smut is caused by ***Ustilago reiliana***.

PEARL MILLET (Vern. Bajra) (Cumbo, Sajja, Cattail millet)

(Pennisetum typhoideum or P. americana)

Origin and History

The origin and history of bajra is not fully known. It seems very probable, however, that the plant originated either in **India** or **Africa**, as the crop is widely cultivated in the various regions of these two places. Pearl millet is also known as **bulrush millet**. ***Pennisetum typhoideum*** is synonymous to ***P. americana*** and ***P. glaucum***. It is a common rain-fed crop in semi-arid areas of Africa and India.

Many workers believe that the pearl millet originated in western tropical Africa since large number of cultivated and wild forms have been seen to grow there. The pearl millet perhaps spread to East Africa and from there to India about 3,000 years ago (Fig. 3.19). The crop is also cultivated-in Arabia and southern Europs, and more recently it has been introduced in the United States of America, where it is being grown mostly for fodder purposes.

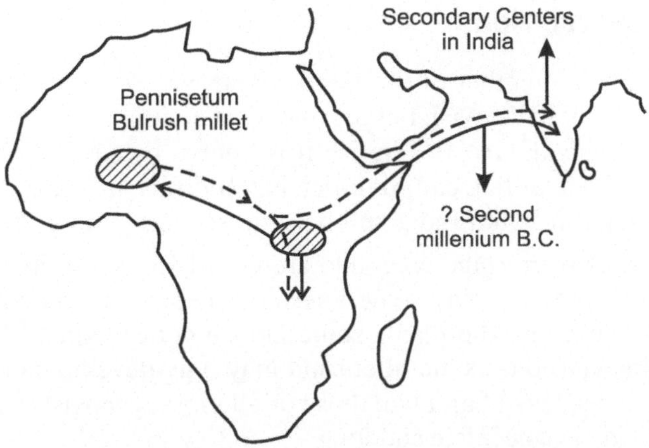

Fig. 3.19. Evolutionary geography of ***P. typhoideum***.

Bajra Growing Regions

World–Bajra is an important cereal crop of tropical Africa, India, Pakistan and South East Asia. In U.S.A. and Europe, it is grown chiefly as a fodder crop. In Sudan its cultivation is next to sorghums.

India—In India, the crop is grown in most parts of the country except Assam. About 80 percent of the annual acreage and 75 per cent of the production is accounted for by the four states—Maharashtra, Rajasthan, Uttar Pradesh and Punjab, The average annual area under bajra cultivation is about 9.48 million hectares and the annual production is 8 million tons.

Fig. 3.20. A photograph of bajra crop.

Botanical Description

Bajra is also a member of the family **Poaceae**. Several scientific names had been given to it in the past, but the one most commonly adopted at present is *Pennisetum typhoideum.* In English it is known by several names such as pearl millet, bulrush millet, cattail millet, etc. In India, it is called by the name of cumbo or sajja in South India and bajra in several parts of North India.

The stems of bajra stand erect and are usually 3 to 8 ft. in height. Tillers are not common but they may sometimes arise depending upon the variety as well as on the spacing. The plants sometimes produce axillary branches, but these are not as vigorous as the tillers and may only develop on two and three internodes. The leaves of bajra like those of all grasses consist of a blade and a sheath. The blade is lanceolate and long pointed and possesses numerous hairs on both surfaces. The sheath which is little thicker than the lamina clasps the stem. The ligule which is a structure at the juncture of blade and the sheath and occasionally known as the rainguard, is about 4 to 5 mm in length.

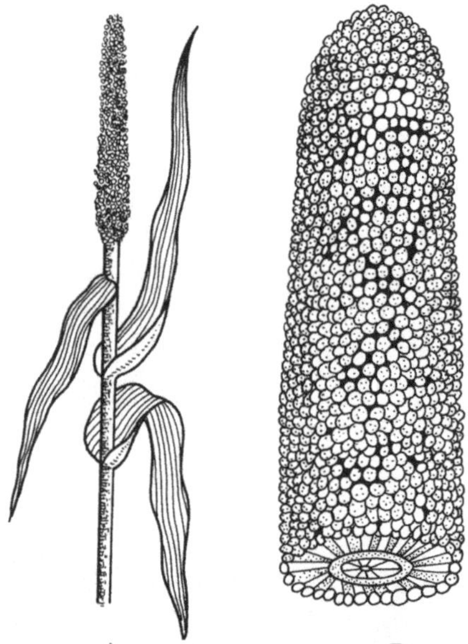

A B

Fig. 3.21. Pearl millet. A. Inflorescence; B. A portion of the axis with grains.

The inflorescence of bajra is a compound spike with an unbranched tapering axis (Fig. 3.21). The length of the inflorescence is usually six to fourteen inches, except in one variety, known as Jamnagar Giant whose inflorescence is about 2 feet or more in length. The branches on which the flowers are borne, known as rachilla, may bear two to five spikelets each but two is the most common number. The inflorescence bears two types of flowers, hermaphrodite and staminate. The former are the first to appear. Soon after the emergence of the inflorescene (head) the styles begin to protrude out of glumes. Cross-pollination takes place in this crop.

The grains are small about 3 to 4 mm long and is 3 to 10 mg in weight. The colour of the grain varies from pale yellow to dull light blue in colour.

The root system like that of most grasses is fibrous. The roots which go downwards are generally thicker than those which spread laterally. On the whole the plant is shallow-rooted and may be considered as a surface feeder. Some high-yielding varieties of bajra are HB-1, HB-2, and HB-4, BJ-104, BE 763, CJ 104, etc.

Ecological Factors

Climate and Soil–Bajra is grown in India as a **kharif** (monsoon) crop but more usually as a catch crop i.e., it is grown when the amount of rainfall during the kharif season has not been sufficient for other crops like jowar, etc. Bajra is suited to warm areas and is resistant to drought. It grows in low rainfall (15 in. to 20 in.) and is best under conditions of light showers followed by bright sunshine. Heavy rains are harmful to the crop. It is grown on poorer types of soil. It can grow well in soils which are too poor for most of the other cereals. The crop is well adapted to sandy or sandy loam soils of Rajasthan where rainfall ranges between 5 in.to 20 in.. Jaipur, Jodhpur and Bikaner divisions account for 28% of the area under cultivation in India.

Cultivation

The preparation of the land on which this millet is sown is not as thorough as that of jowar. The land is usually ploughed two or three times before sowing. The seed is usually sown mixed with other crops preferably pulses, *viz.,* pigeon pea, urd, mung and seasamum, etc. The seeds are sown broadcast. The crop is usually not weeded and irrigation is not done. The crop is seldom manured. Irrigated crop may be manured more liberally.

Harvesting

The crop matures in October. The crop, when grown for grain, is harvested when the heads are ripe; but when grown for fodder, it is cut before the stalks become too dry. The plants are cut by means of sickles as near the ground as possible and the heads are taken to the threshing floor where the bullocks tread the seed out. The grains are cleaned by winnowing and are then stored in the mud pots.

Uses of Bajra

1. Grains are ground into flour for preparing bread.
2. It can be fried for producing a nutritious food.
3. The stalks are fed to cattle and are also used for thatching purposes.

Diseases

The most common diseases which occur on bajra in our country are:

1. 'Green ear' disease, This is caused by fungus known as ***Sclerospora graminicola***. The disease, as the name indicates, usually appears in the form of a mass of small twisted green leaves in place of the grains in the spike or head.
2. Rust is caused by ***Puccinia penniseti.***
3. Smut is caused by ***Tolyposporium penicillaria.***

RAGI
(Finger Millet)

(Eleusine coracana)

Origin and History

Ragi is also known as **African millet** (Koracan). The finger millet is believed to have originated either in India or Africa. It has best storage properties of all the millets and can survive for as long as 5 years in the unthreshed form.

It is one of the ancient '**minor millets**' of India. It is closely related to the common weed, goose grass, ***Eleusine indica*** from which it is believed to have originated. Since it is a tetraploid it might be a true breeding hybrid of latter and, some other species, Six of nine species of ***Eleusine*** grow in the **tropical Africa**. While ***E. coracana*** is a tetraploid (2n = 4x = 36), ***E. indica***, a pantropical species is diploid (2n=18). The cultivated ***E. coracana*** has two types which resemble the two subspecies of ***E. indica*** viz. ***E. indica*** subsp. ***africana*** and ***E. indica*** subsp. ***indica***. It is believed by many that cultivation of ***E. coracana*** originated from the subsp. ***africana*** in **Uganda** or an adjoining region from where it spread to India about 3,000 years ago at about the time when the bulrush millets were also taken to India (Fig. 3.19).

Ragi Growing Regions

World–It is cultivated in many parts of the world for grain and forage. Ragi is grown in several countries of Africa like Uganda, Nigeria, Zambia and in Sri Lanka, India, Malaya, China and Japan.

India–Among minor millets Ragi is most important in India. The largest acreages are found in Karnataka (30% of the total) and Tamil Nadu but it is also grown in Maharashtra, Andhra Pradesh, Uttar Pradesh and Bihar. The area under cultivation is about 3 million hectares.

Botanical Description

The plant is a tall tufted grass which attain a height of 4 feet. It has flattened stems which are enclosed by numerous leaf-bases. The main stem and its tillers produce a very much branched shallow fibrous root system. Each stem bears four to six dense, digitate spike inflorescences at the tip (Fig. 3.22). The spikes are 4 to 5 inches in length and bear numerous spikelets. The spikes may be stiff, erect, divergent or curved. The fruit is not a typical caryopsis. The pericarp is separate from the seed. The seeds are small and rounded. They are dull red in colour.

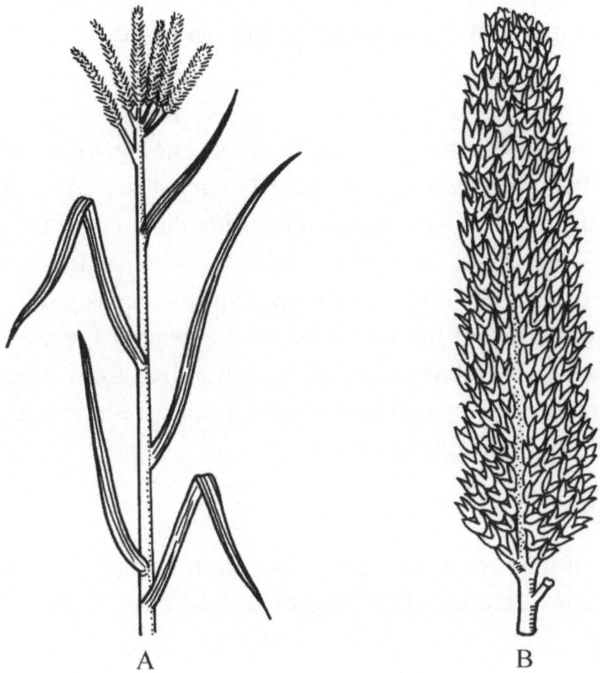

Fig. 3.22. *E. coracona*. A. A portion of plant showing digitate inflorescence; B. A single inflorescernce.

Ecological Factors

Ragi is one of the hardiest crops. It can be cultivated under dry conditions. The rainfall requirement ranges between 20 and 40 inches. It can be cultivated as an irrigated crop in the hot season. In south India, the crop is grown mainly on red, light red, light black or ashy loams and sandy loams. Ragi can grow very well even on poor soils. In some parts of India like Uttar Pradesh, Gujrat and Bihar it is cultivated on well-drained alluvial loams. It can tolerate a certain, degree of alkalinity in the soil. Ragi responds to high fertilization and large yields are obtained under favourable conditions.

Cultivation

The rainfed crop is grown from May to August or July to November and the irrigated crop all the year round particularly in Karnataka and Tamil Nadu.

A clean and smooth bed is needed for good growth. The land is prepared by ploughing after the harvest of the previous crop followed by another two or three ploughings after the start of rains.

Ragi is sown in lines or broadcast. Four to five week old seedlings are usually transplanted in lines which are about ten inches apart.

Harvesting

The crop matures in three or five months. The plants are cut close to the ground, tied into bundles which are stacked to dry. The grain is threshed out by either beating the bundles with sticks or treading them under the feet of oxen or by working a stone roller over them. In some parts of India, particularly in Tamil Nadu, the irrigated crop does not ripen simultaneously. They are collected as they mature, carted to the threshing floor, heaped for a few days to get 'cured' and then threshed as usual. The grain is free from insects and can be stored for a long time. New and improved varieties are being developed by the Indian Agriculture Research Institues.

Uses of Ragi

1. Ragi is an important food crop. The grain is considered nutritious. It is usually converted into flour and made into cake or pudding and a thick porridge.
2. Malt, i.e., fermented beverage is made from the grain.

KODO MILLET
(Vern. Kodon, Pakodi Arika)

(Paspalum scrobiculatum)

The kodo millet is one of the minor cereals. It is widely cultivated in our country in the southern states. The small grains are used for food by the poor people. The grains are very coarse. The husks and immature grains are poisonous.

The crop is grown in rainfed regions and is highly drought resistant. The crop is sown just after the onset of monsoon in late June.

Botanical Description

The plant belongs to **Poaceae** family. It is an erect annual crop. It is 50 to 75 cm in height. The internodes are solid. The leaves are linear or lanceolate. The ligule is membranous. The plant produces large number of tiller (up to 18). The inflorescence is a panicle of two to eight spikes. Spikelets are arranged alternately in two series. The plant is self-pollinated. The grain is fully ripe after thirtyfive days of flowering. The grain is a one-seeded caryopsis.

THE COMMON MILLET

(Proso Millet; Vern. Vari, Kutki)

(Panicum miliaceum)

The common millet is known as Indian millet, Broomcorn millet, Hog millet, Panicum millet and Russian millet. It is grown for food of man and livestock.

They grow very rapidly and more quickly than any other cereals.

Origin and History

Common millet is known to have been domesticated in the **southern Europe** and **Asia** in the prehistoric times. It was the 'millium' of the Romans. It finds reference in the Old Testament. It is believed to have originated in **India**, **Egypt** or **Arabia**. It is the true millet of history.

GRowing Regions

World–It is grown widely in south eastern Russia, China, Japan, India, southern Europe, and parts of North America.

India–In India it is widely grown is Tamil Nadu, Andhra Pradesh, Karnataka, Madhya Pradesh and Maharashtra.

Botanical Description

The plant belongs to the **Poaceae**. It is an erect annual. It is three to four feet in height. It has a shallow adventitious root system. The internodes are hollow. The branching is sparse. Tillers are very weakly developed. The leaves are narrow and pointed. The inflorescence is a compact panicle (Fig. 3.23). A spikelet contains one sterile and one fertile flower. The latter has two lodicules, three stamens and an ovary with two long styles and feathery stigmas. Flowers are either self or cross-pollinated. The grains are variously coloured.

The grain is highly nutritious. It contains plenty of carbohydrate, 10 per cent protein and 4 per cent fat.

Fig. 3.23. Common millet. A. Plant; B. Ripe ear,

Ecological Factors

Of all the grain crops common millet has the lowest requirement of water. It grows in conditions where no other crop can survive. It grows in the rainfed regions of the hot tropics and subtropics where the soil conditions are extremely poor.

Cultivation and Harvesting

The field is ploughed two or three times usually after the first rain. The ploughings are followed by two or three harrowings by tooth harrow. The seeds are sown by broadcast, drilling or transplanting method during the month of July.

The crop ripens very quickly in only three months. The crop harvested by cutting the stalks near the ground level which are taken to the threshing floor. Threshing is done by making the bullocks to tread upon the dry grains. Threshing is followed by winnowing.

Uses of Common Millet

1. It is poor man's food grain. Grains are eaten as rice.
2. The flour of the millet is used in bread making.
3. The flour is also used for preparing porridge and pudding.
4. It is used as food of the livestock.
5. It is a source of starch in Russia.
6. It is used as bird-seed in Britain.
7. The straw is used as fodder.

LITTLE MILLET

(Panicum miliare)

It has very small sized grains. It is grown in India, Sri Lanka, Pakistan, Burma, and South-East Asian countries. The crop is limited to the warm tropics and subtropics. It can also be cultivated up to an altitude of 7000 feet. The crop is mainly cultivated in India. The states of Uttar Pradesh and Madhya Pradesh are the chief growing areas of the little millet in our country.

The plant can grow on extremely poor soils. The crop is highly drought resistant and is also capable of withstanding water-logging,

Botanical Description

The plant is an erect annual. It has a height of 0.75 to 1 meter The stem bears linear leaves. The plant produces tillers. It. bears a highly branched panicle (Fig. 3.24). The spikelets contain a complete flower and a staminate or sterile flower. Self-pollination is common in the flowers. The grain is a caryopsis fruit. It is olive brown or light olive brown.

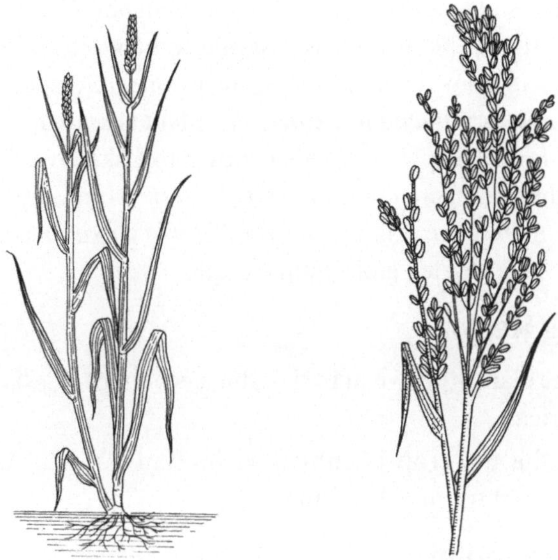

A B

Fig. 3.24. Little millet. A. Plant; B. Ripe ear.

FOXTAIL MILLET
(Italian Millet; Vern. Rala, Kangni)
(Setaria italica)

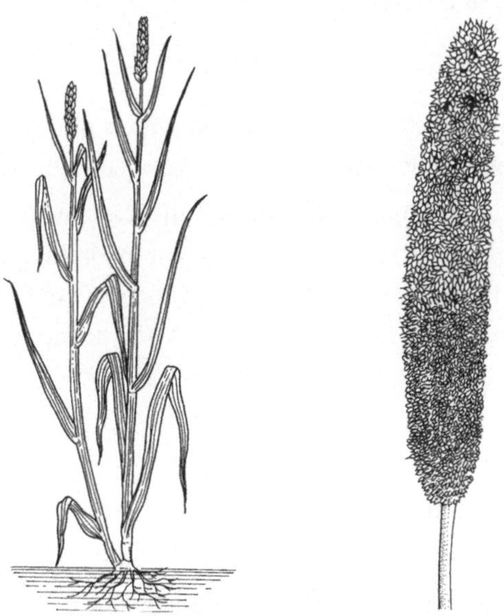

A B

Fig. 3.25. Foxtail Millet. A. Portion of a plant; B. Inflorescence.

Foxtail millet is also known as Italian, German and Hungarian millet. The place of origin of this millet is not clear. While some workers believe it to be of Italian origin others consider it to be Asian in origin. According to **Vavilov** (1935) Foxtail millet originated in **China**. The plant is known to be sacred to the Chinese in 2700 B.C. The plant was also known to Lake Dwellers in Europe. It is believed that the plant spread to India and Europe from Central Asia (China). The plant is thought to have originated from *Setaria viridis,* a close wild relative which occurs in Asia, Africa and South Europe.

Rala Growing Regions

World—The plant is grown in India, China, southern Russia, eastern Europe and North America.

India–In India the crop is cultivated in Tamil Nadu, Andhra Pradesh, Karnataka. Maharashtra and Haryana.

Botanical Description

The plant belongs to the family **Poaceae**. It is ah annual herb having a height of 1 to 1½ metres. The slender stem has jointed internodes. The plant bears narrow linear leaves. It bears a terminal cylindrical inflorescence (Fig. 3.25). It is about one foot in length. The inflorescence is a spike and is usually compact in nature. The fruit is a caryopsis. The grains are white, yellow, brown, black or red in colour.

Ecological Factors and Cultivation

The plant requires 500–750 mm of rain which may be limited to only two months. It can grow on poor soils but gives best yield on loamy soils.

The field is first ploughed which is followed by two to four crosswise harrowings with a blade harrow. Seeds are sown by drilling or broadcast method during the rainy season in the northern India and during September-October in the South. Seeds are sown from January to April in the irrigated areas.

Uses of Foxtail Millet

1. It is poor man's food.
2. It is grown for fodder particularly in U.S.A.
3. The grains are used as bird feed.
4. It is used in preparing porridge and custards.

JAPANESE BARNYARD MILLET

(Echinochloa colona var. *frumentaceae)*

The Japanese Barnyard millet is believed to have originated from ***E. crusgalli*** which occurs as an annual weed in rice fields and wet areas of many parts of **Africa**, **Asia** and **America**. The millet is grown in India, China, Japan, Korea, Malaysia, Indonesia, in some parts of Africa and the United States. In our country the crop is grown mostly in the southern states. The plant is grown for food as well as fodder. The crop grows very quickly and ripens in just 45 days.

Botanical Description

The plant is an annual herb. It grows to a height of four feet. It has few tillers. The slender stem bears few leaves. The inflorescence is an elongated panicle which, has several (up to fifteen) branches (Fig. 3.26). Each spikelet has one sterile and one perfect flower.

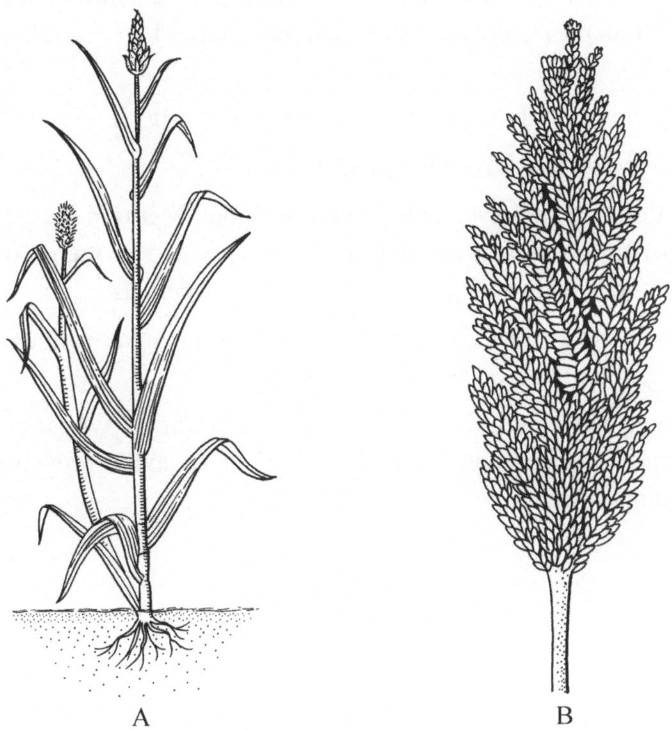

Fig. 3.26. Japanese Millet. A. Portion of a plant; B. Inflorescence.

It is grown in warm regions as a rainfed crop. The crop can also be cultivated on submerged lands on the banks of rivers.

Uses of Japanese Millet

1. The millet is eaten along with rice in the form of a porridge. The grain is ground into meal for preparing the porridge.
2. It is grown for fodder in the United States.

QUESTIONS

1. Which is the chief cereal crop of the world? Mention the botanical features of the plant and state the ecological factors suitable for the crop.
2. Describe the distribution, botanical features of a cereal which has originated from New World.
3. What do you know of the breeding programme carried on maize crop?
4. Which is the chief millet crop of India? Give a detailed account of the soil, temperature and rainfall conditions required for the cultivation of the crop.
5. Write, notes on the following regarding their origin and distribution.

   (i) Rice, *(ii)* Barley and *(iii)* Oat
6. What are the major cereals grown in India? Mention the origin (native home) and botanical characteristics' of any one of them. What are the advantages of growing 'dwarf' varieties of cereals?
7. What is meant by 'Green Revolution? Discuss the role of genetics in bringing about green revolution in India.
8. What are the cereals? Give a brief account of wheat with special reference to its botany, methods of cultivation and improvement through breeding in India.

4

Food Plants (Contd.)

LEGUMES

Legumes are next only to cereals as sources of food. They constitute an important part of people's diet. They are popularly known as pulses. The beans, peas, soybeans, etc., are sources of very high protein-rich seeds. These vegetable proteins are an essential balance to the carbohydrate-rich cereals.

Pulses grow easily and mature rapidly. They are highly nutritious. They supply not only proteins, but also carbohydrates, fats, vitamins and some minerals. The immature seeds are also used as vegetables. The protein-rich plants are very good forage crops. The roots of these plants possess nodules which contain nitrogen-fixing bacteria. These bacteria fix atmospheric nitrogen and thereby enhance the fertility of the soil.

The legumes owe their name to the fruit-'legume'or pod which is a simple, dry, dehiscent fruit dehiscing from both the sutures. The protein rich seeds possess a very hard seed coat and have low water content. These two qualities enhance their storage value. The plants-bearing legumes belong to the great family **Leguminosae**.

Leguminous plants have been cultivated for centuries all over the world. In India the area under pulses is nearly 23.76 million hectares and the annual production touches over million tons. Due to shortage in pulses the Government has embarked on a plan to produce pulses to the tune of more than 14.11 million tons. Leguminous crops are very commonly grown for crop rotation since on the one hand they are excellent nitrogenous manures, when dead and ploughed under. On the other hand they bring about an increase in the nitrogen content of the soil with the help of the nitrogen-fixing bacteria present in their root nodules.

GRAM

(Chick Pea, Garbanzos; Vern. Chana)

(Cicer arietinum)

Origin and History

Gram or chick peas are also called garbanzos or garbanzo beans. Gram is extensively consumed by both animals and human beings. Gram contains 20-22 per cent of crude protein, 50-60 per cent of carbohydrates and 5 per cent oil.

Radio-carbon-dating of the first remains of gram discovered in **Turkey** has indicated it to be of 5450 B.C. The center of origin of gram is thought to be the **western Asia**. It is believed that the ancestors of ***C. arietinum*** *were* taken to the Mediterranean region in the fourth millenium B.C. and to India by the Aryans at a later date. **Chowdhury et al.** (1970) has reported the discovery of the first remains of gram in India from Atranji Khera in Uttar Pradesh. The remains belong to 2,000 B.C. There are also reports of findings of gram from South India dating back to 300—500 B.C. or second millenium B.C. (dating doubtful). It is believed that gram got introduced in our country by a land route as well as an oversea route.

Gram is believed to have spread to Ethiopia from the Mediterranean region. Gram has been introduced in South and Central America (Mexico) in the sixteenth century. Later Indians introduced it in East and South Africa and the West Indies.

Gram Growing Regions

World–India accounts for 90% of the total world production of gram. The other gram growing regions are around the Mediterranean (Iran, Iraq, Morocco, Algeria, Greece, Italy, Portugal, Spain), Ethiopia and Mexico.

India–In India it is chiefly grown in Punjab, Uttar Pradesh, Bihar, Madhya Pradesh and Andhra Pradesh. Northern India accounts for nearly 90% of the total covered area and about 95% of the total annual production, which are 7.63 million hectares and over 6 million tons respectively.

Botanical Description

Gram belongs to the family ***Leguminosae***. It is a small, herbaceous annual. It is much branched plant and attains a height of about 2 feet or more. The plant may be of erect or a bushy spreading type. A variety of gram known as 'Kabuli chana' is generally taller than the local Indian crop and stands more or less erect.

The leaves are pinnately compound, usually with one terminal leaflet (Fig. 4.1). The number as well as the size of leaflets, however, vary in different varieties.

Fig. 4.1. A portion of a chick pea plant.

The flowers are papilionaceous, and possess various colours, which vary from white or greenish to various shades of pink or blue. The flowers are usually solitary and are developed at a node opposite to leaf. The pods usually contain two seeds. Flowers are normally self-pollinated, nevertheless natural crossing occurs occasionally. Seeds vary in size as well as in colour. The seeds of certain types are whitish while others are either yellowish, brown or reddish. Gram has a well-developed root system. There is a central tap root system with numerous lateral branches which spread out in all directions.

Ecological Factors

Climate–It is well adapted to arid and semi-arid regions. It is a cool season (October-March) crop in our country. It is best suited to areas having low to moderate temperature and a mild cold season. Gram is fairly sensitive to the conditions of the weather. Heavy showers soon after sowing or at flowering and fruiting stages cause great harm. Frost during this period will result in the failure of the flowers to develop seeds or in the killing of the seeds inside the pod. Hailstorms at the ripening time cause great loss. Early summer heat shortens the growing period, hastens maturity and reduces yield.

Soil–The best kind of soil for gram is one which is well drained and is not too heavy. Some of the heavier soils which are good for wheat are not very suitable for gram. So in most parts of the country gram is grown in fairly light soils, mostly sandy loams, which are too poor for wheat. ***Kabuli*** gram, however, requires better soil. In Maharashtra and south India, it is cultivated on water-retentive clay loams and black cotton soils.

Cultivation

The land where gram is to be sown usually does not require as good preparation as in the case of wheat. Generally one or two ploughing are given and ordinarily the seed bed is pulverized. Gram is usually sown earlier than any other **rabi** crop. Sowing is usually done about the beginning of October in most parts of northern India. The crop does not require any care after sowing. The crop is usually not irrigated. It is believed by most cultivators that nipping off the tips of the shoots just before flowering increases the yield. The crop is rarely manured.

Harvesting

When the leaves of the crop begin to turn yellow and the plants begin to dry up, the crop is ready for harvesting. Plants are cut with a sickle and are then taken to the threshing floor. The cut plants are allowed to dry for a few days, and are then threshed by allowing the bullocks to tread upon them as in the case of wheat. The chaff ('bhusa') is separated by winnowing.

Uses of Gram

1. Gram seeds (cotyledons) are boiled and cooked as a pulse.
2. The gram flour is very nutritious and used by poor people for making breads.
3. 'Besan', a flour like preparation of the whole grain is used for preparing 'Pakoras' etc.
4. The dry grains of gram are eaten as such or are roasted, or boiled and salted for use as light refreshment by the poor people. They are also cooked as vegetable. Boiled chick peas taste like chicken.
5. The green seeds are sometimes eaten raw.
6. The grains are very commonly used for feeding horses etc.
7. The chaff ('bhusa') removed after winnowing is used as cattle feed.
8. Seeds are used as a substitute for and as an adulterant of coffee.
9. Gram flour is used as a substitute for wheat flour for diabetic patients.

Diseases

The gram crop is susceptible to fungal infection. Rust of gram is caused by ***Uromyces ciceris arietini***.

PEAS
(Vern. Mattar)
(Pisum sativum)

Origin and History

The well-known pea derives its English name from the Latin word ***Pisum*** (pee sum). The English first turned **pisum** into **pease**, and then shortened the word to **Pea**.

The pea is probably indigenous to **southeastern Europe**, and **western Asia**. Burned peas have been found in the ruins of lake dwellings in Switzerland and in the ancient city of Troy, which was first built during the Stone Age. Archaeological findings of carbonized pea seeds in West Asia and Europe have been dated to 7000 B.C. According to **Vavilov** (1949) **Ethiopia**, the **Mediterranean** and **Central Asia** with a secondary centre in **West** are the centres of origin of pea. The exact primary centre of origin is, however, not known.

Peas of early times were small, and were dried before cooking. They belonged to the variety now called field peas. The field peas ***(Pisum arvense)*** are believed to have originated from gray pea, which still grows wild in Greece and the Levant. The field peas are grown even today. Their flowers are coloured and the seeds are green or yellow.

Garden peas ***(Pisum sativum)*** are larger and sweeter than the field peas, and are of a recent origin (less than a thousand years ago). They were first sold in England in 1100's. Spaniards took peas to Mexico before 1540. Garden peas have white flowers and round, smooth or wrinkled seeds. The seeds are white or yellow in colour. Gregor Mendel carried experiments on garden peas.

While field peas are grown as a field crop the garden peas are grown as vegetables.

Pea Growing Regions

World–The principal pea growing countries are China, U.S.A., India, Canada and Northern Europe.

India–In India it is commonly grown as a field crop in Uttar Pradesh, Bihar, Madhya Pradesh, Himachal Pradesh, Punjab, Orissa and Maharashtra.

Botanical Description

Pea plants are semi-erect, climbing, or trailing annuals. The stems grow to a height of about 2 to 3 feet. The leaves are pinnately compound with 2 to 6 leaflets in pairs (Fig. 4.2): The leaves terminate in tendrils. The stipules are large and leaf like. The inflorescence is an axillary raceme. Flowers are papilionaceous with a diadelphous and roccium. Self-pollination is common, but cross-pollination may also occur. The fruits are legumes or pods which are simple, dry and dehiscent.

A B

Fig. 4.2. A. A portion of a garden pea plant showing leaflets, stipules, tendril and pods; B. An opened pod.

The flowers and seeds of field peas and garden peas are different from one another:

	Field pea (Choti matar)	**Garden pea** (Bari matar)
Flower	coloured	white
Seed	small	large
	smooth	smooth or wrinkled,
	angled	round,
	green or yellow	yellow or white,
	eaten as pulse	eaten as such,
	not very sweet	sweeter.

Ecological Factors and Cultivation

The pea plant is adapted to a moderately cool growing season and a fair amount of rainfall. It is generally grown on heavy soils such as loams and clay loams.

The soil is prepared as in the case of gram and other **rabi** crops. Seeds are sown broadcast or with the help of a grain drill. The seeds are then covered with soil by planking or harrowing. If sown in October, the crop is usually ready for harvesting in the month of March or April. If the crop is grown for seeds harvesting is done when the seeds are mature and dry. Dry peas are easily cut and are removed to the threshing floor where they are threshed out by bullocks. A threshing machine called '**Viner**' has been invented in U.S.A. It is capable of removing green peas from pods. This has resulted in a large processing industry.

Uses of Pea

1. Seeds of field peas are consumed by human beings in the form of pea meal or split peas.
2. The seeds are consumed by the livestock.
3. The plants are used for forage, silage and as green manure.
4. The seeds of garden pea are eaten green or are used as vegetables.
5. They are also canned and frozen by the canning industry. Pre-cannery refuse is useful for feeding the livestock.

Diseases

Pea like other legumes is infected by certain fungi which cause considerable harm to the plant. The most common diseases are

1. Downy mildew caused by ***Peronospora pisi***.
2. Powdery mildew caused by ***Erysiphe polygoni***.
3. Rust caused by ***Uromyces fabae***.

PIGEON PEA

(Cajan pea; Vern. Arhar)

(Cajanus sp.)

Origin and History

Pigeon pea is a hardy pulse of tropical grassland and arid regions. It accounts for about four per cent of the total pulse production in the world.

Botanists differ in their opinion regarding the place of origin of the pigeon pea. Some botanists think Asia, particularly India, to be the place of its origin. The majority, however, favours Africa to be the centre of origin. Since the plant has been found wild in parts of Africa, it seems to be the more probable place of origin. The plant, therefore, might have been brought to this country by the early traders who carried on trade between India and Africa. It is now one of the most widely cultivated pulses in India and accounts for 95 per cent of its world production. **D.N. De** (1974) has provided strong evidences to prove that pigeon pea has originated in the peninsular India. Two types of ***Cajanus***, Tur ***(flavus)*** and Arhar ***(bicolor)***, have Indian names derived from Sanskrit words indicating their evolution at a very early date. The crop spread to Malaysia from India around 2000 B.C., to China around 1500 B.C:, to East Africa (Zanzibar) between 1500 and 1000 B.C. The crop then spread to West Africa and to the Nile Valley from East Africa. Portuguese carried the crop to the New World from Zaire or Angola in early post-Columbian days.

Cajanus cajan is believed to have arisen from an unidentified species of its closely related genus ***Atylosia***.

Arhar Growing Regions

World–At present it is grown in almost all the tropical and subtropical regions of the world including Uganda, America, Italy, Australia, Pakistan, West Indies, India and South-east Asia.

India–In this country the pigeon pea is usually grown as a mixed crop with jowar, bajra, cotton and many other crops. It is mainly grown in Bihar, Madhya Pradesh, Bengal, Uttar Pradesh, Assam, Maharashtra and Tamil Nadu. It is grown in 3.53 MH and the yield is 2.51 MT.

Botanical Description

The pigeon pea, like all pulses, is a member of the family **Leguminosae**. Two varieties of this crop are generally recognized–the late maturing variety and the early maturing variety. The plant is an erect shrub. It is always cultivated as an annual but if allowed to remain in the soil lives for three or four years. The height as well as the circumference of the plant depends on the fertility of the soil and the amount of space given to it. It may attain a height of 6 to 7 feet. The plant is very much branched. The leaves are trifoliate. The leaflets are entire and densely-silky on the lower surface. The total length, shape and size of leaf and leaflets vary with the varieties.

The inflorescence is an axillary raceme (Fig. 4.3A). The flowers are distinctly papilionaceous. In the late varieties the flowers, are usually grouped together at the end of the branches, but in the early varieties the flowers are produced at several points along the branches. The process of flowering continues in each plant up to the time of harvest. The flowers are mostly self-pollinated.

A B

Fig. 4.3. A. A flowering branch of the pigeon pea; B. Fruits

The fruit of the pigeon pea is a pod which is usually constricted obliquely between seeds (Fig. 4.3B). When the pods are constricted they are termed '**bead**' otherwise they are termed '**not bead**'. The pods vary in colour from green to dark. The seeds within the pod vary in number. They also vary a great deal in colour, the majority being either brown or white but they may also be purplish black or dark mottled brown.

The root system of the pigeon pea consists of a central tap root with numerous laterals and secondary branches. The strong and deep tap root makes the plant remarkably drought resistant. One of the reasons why the pigeon pea is commonly used in this country as a renovating crop is its deep and penetrating roots which break up the subsoil, inducing a better physical condition.

Climate–The crop usually prefers a fairly moist climate during the period of its vegetative development. It is for this reason that it is commonly sown with **kharif** crop. Bright sunny weather during the flowering and ripening stages is needed for the copious setting of fruits. It is drought resistant. One of the main factors which limits its growth in the northern part of India is its ability to withstand frost.

Ecological Factors

Soil–The crop may be grown in all kinds of soils, but it prefers a light and moist soil in which the roots can develop freely. The soil should not be deficient in lime. In India, the soils in which the crop grows ranges from the alluvial soils of Indo-Gangetic plains to heavy black cotton soils of Maharashtra and central and southern India. However, it grows best on well drained, light to medium soils.

Cultivation

The preparation of land, when, the pigeon pea is sown mixed with other crops, is the same as that for the crop sown with it. If this crop is sown alone the land is ploughed once and harrowed two or three times to obtain fairly pulverised seed beds.

The crop is sown broadcast or in rows, whereas the mixed crop is almost always sown in lines, one row of pigeon pea being planted after every four to ten rows of the main crop of jowar, bajra or cotton, etc.

The crop is seldom weeded. Pigeon pea usually does not need any irrigation, but sometimes water is given in order to protect the crop from damage by frost.

Harvesting

When the plants are ripe, the stalks are cut as near the ground as possible by means of sickles. They are then tied in bundles and taken to the threshing floor and allowed to dry for a few days. The leaves and pods are then stripped off from the stems and heaped up in a pile. These are usually threshed by allowing bullocks to tread on them.

Uses of Arhar

1. Pigeon pea or red gram is extensively used as dal.
2. Green pods are used as vegetable.
3. The leaves and broken pods are fed to cattle while the stalks are used for roofing, basket making etc.
4. The pigeon pea rivals alfalfa as a forage plant.

Diseases

The most common disease which attacks the pigeon pea is known as wilt and is caused by ***Fusarium oxysporium***.

LENTIL

(Vern. Masur)

(Lens culinaris syn. L. esculenta)

Origin and History

The lentil is one of the most ancient legumes of the **Old World**. The carbonized remains of the seeds of lentil discovered in **West Asia** date back to 7000–6000 B.C. Later lentils spread to Greece and Bulgaria. Along with wheat and barley lentils spread to the central Europe, the Mediterranean in the Neolithic and Bronze-Age agricultures. Lentils have been found in the ruins of the Stone Age lake dwellings as well as in Egyptian tombs which are more than 4000 years old.

Lens culinaris is believed to have originated from ***Lens orientalis*** which is distributed in Turkey, Syria, Israel, Northern Iraq and western and northern Iran (Fig. 4.4).

Lentil Growing Regions

World–The area of lentil cultivation embraces several countries of Asia, Africa and United States of America. India, Pakistan, Ethiopia, West Asia, the Mediterranean region are the chief lentil producing areas of the world.

India–In India it is cultivated in many parts of the country. It is grown throughout the north India, Madhya Pradesh and in some parts of Maharashtra. It is of very little importance in South India. Lentil is grown in about 1.50 MH and the total yield is 0.95 MT.

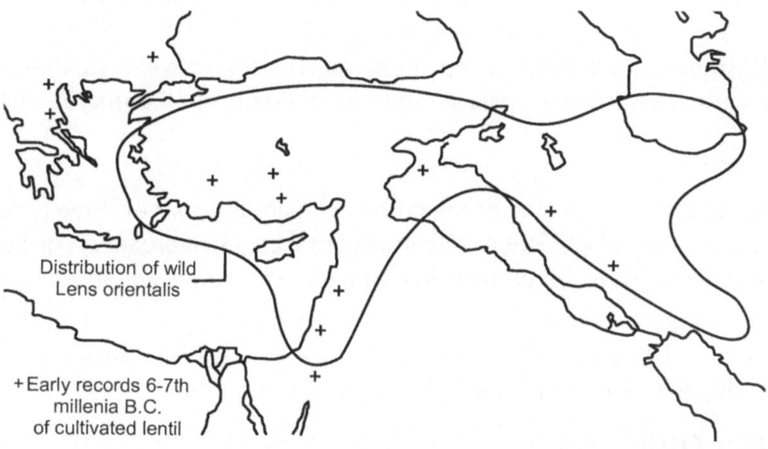

Fig. 4.4. Distribution of wild ***Lens orientalis*** and early records of cultivated plants.

Botanical Description

Lentils are slender, much-branched annual herbaceous plants. They vary in height from about 6 to 24 inches. They are usually more or less erect. The general habit of the plants depends upon the mode of branching, the height and also on the amount of foliage. The stems are usually angular and possess short hairs, the degree of hairiness varying with different varieties. The stems may be entirely green or may possess anthocyanin pigments at the base. There are several varieties of ***Lens culinarias*** grouped under two subspecies *viz.* subsp. ***microsperma*** having small seeds and small pods and subsp. ***macrosperma*** having large seeds and large pods.

The leaves are pinnately compound with 2 to 8 pairs of oval or linear leaflets. They terminate in tendrils (Fig. 4.5). The leaves are stipulate. The inflorescence is raceme. The flowers may be white, pink or violet. The seeds are thin and flat. The cotyledons ('dal') are pink coloured.

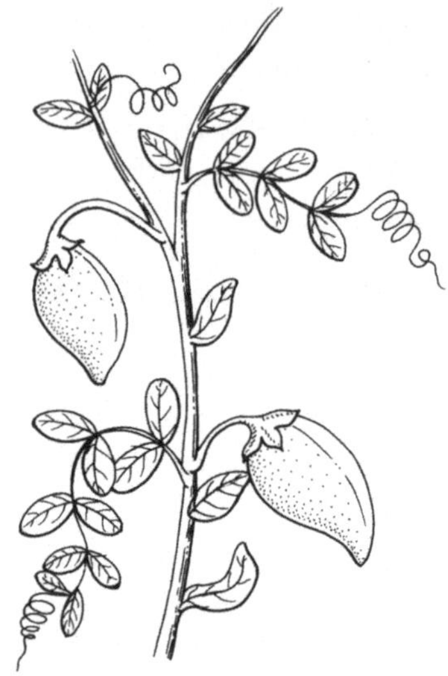

Fig. 4.5. A portion of a lentil plant.

Ecological Factors

Climate–In our country it is generally grown as a winter crop. It can be grown even at an altitude of 11500 ft. in the extremely cold conditions. It is generally grown as unirrigated crop but when sown late it may require one irrigation.

Soil–The lentil crop is grown over a very wide range of soils. The crop is raised on light loams and alluvial soils in northern India and on well drained moderately deep, light black soils in Madhya Pradesh and Maharashtra. The crop can bear a moderate amount of alkalinity of the soil.

Cultivation

The seed bed for sowing of the crop is prepared by two or three ploughings which are considered sufficient. The seeds are then sown broadcast or in rows 9 to 12 inches apart. It is sometimes grown mixed with barley and mustard or both.

The crop generally takes 3 to 4 months for maturity. It is harvested before it is fully ripe. The cut plants are dried for a few days on the threshing floor, and are then threshed and winnowed like other pulses.

Uses of Lentil

1. As a pulse it ranks equal to pigeon pea and is inferior to black and green grams in nutrition. The French and Spaniards cook brown lentils for soup and grind yellow one into flour.
2. Unripe pods are used as green vegetable.
3. The seeds are more digestible than meat and constitute excellent substitute for meat in many Catholic countries.
4. Dry leaves, stalks, husk and broken grains are used as cattle feed.
5. In some parts of our country it is grown for green manuring of paddy.

Diseases

1. Dry root rot is caused by ***Macrophomina phaseoli***.
2. Rust is caused by ***Uromyces fabae***.
3. Wilt is caused by ***Fusarium orthoceras*** var. ***entidis***.

MUNG BEAN OR GREEN GRAM OR GOLDEN GRAM

(Vern. Mung)

(Phaseolus aureus)

Origin and History

It is one of the common pulses in India, where it has been cultivated since ancient times. Mung or green gram is probably indigenous to **India**. Its cultivation in Africa is more recent. It is a very important crop of our country. Mung is also grown in China and some other Asiatic countries.

Its cultivation in India is widespread. The most important Mung producing states are Madhya Pradesh, Maharashtra, Uttar Pradesh, Punjab, Andhra Pradesh, Rajasthan, Karnataka and Tamil Nadu.

Botanical Description

Mung belongs to the family ***Leguminosae***. Mung is a small herbaceous annual growing to a height of about 1 to 3 feet. The main stem is more or less erect. The lateral branches are sub-erect. The leaves are trifoliate, the leaflets being large, ovate and entire. Both the stem and the leaves are covered with short hairs. The flowers possess various shades of yellow colour and are produced in a cluster. The pods are slender and about 3 to 4 inches long. The seeds are small and nearly globular. The colour of the seeds is usually green. The cotyledons ('dal') are yellow in colour.

I.A.R.I. has developed several varieties of Mung which can be intercropped with maize, sorghum and bajra without affecting productivity of anyone of them.

The Mung bean varieties–PS-10, PS-7 and PS-16 are specially suitable for cultivation during summer months. The variety S-8 gives good yield during the **Kharif** season.

Ecological Factors

Climate–The plant requires a well distributed rainfall of 25 to 35 inches. Heavy showers at flowering time cause considerable damage. Even damp winds at the flowering stage affect the fertilization. Mung is generally grown in this country as a **kharif** crop but in some parts of the country such as Maharashtra, it is grown as a winter crop.

Soil–The soil on which it is cultivated are usually loams which are deep and well-drained. It is also cultivated on light or shallow strong soils as well as on clayey soils.

Cultivation

The land is ploughed once or twice and is then harrowed. The kharif crop is sown in June or July and the rabi crop in September or October. The seed bed preparation is similar to that of other pulses. The seeds may be sown broadcast and covered by planking or they may be drilled in rows 9 to 12 inches apart. There is no need for irrigating the crop since it is grown in monsoon season. Flowering starts in about 60 days and the crop matures after three or four weeks of flowering.

The crop is harvested before it is fully ripe as there is a danger of the shattering of the dry pods. The plants are cut by means of sickles and are removed to the threshing floor and dried. The seeds are then threshed by means of bullocks and are cleaned by winnowing.

Uses of Mung

1. The green pods are used as vegetable.
2. Seeds (whole or split) are used as pulse.
3. The seeds are easily digestible, without producing heaviness. Grains are eaten whole, parched, salted, with sugar, or are boiled with condiments.
4. The Chinese consume the germinating 'bean sprouts' when they are about four days old.
5. The crop also helps in restoring the fertility of the soil.
6. Decoctions of the seeds are also reportedly used medicinally in treatment of colic pain.
7. Straw and husk are used as fodder for cattle.

Diseases

1. Leaf spot is caused by ***Cercospora cruenta***.
2. Powdery mildew is caused by ***Erysiphe polygoni***.

URD BEAN

(Black Gram; Vern Urd)

(Phaseolus mungo)

Urd Growing Regions

Urd is a common pulse of India. It is believed to be a native of India. It is cultivated in many tropical and sub-tropical countries of the world. It is grown in several regions in **Africa**, **southern countries of Europe** and **India**.

It is grown all over India, the main areas of production being Madhya Pradesh, Uttar Pradesh, Punjab, Maharashtra, Bengal, Andhra Pradesh and Karnataka.

Botanical Description

Urd belongs to the family **Fabaceae**. The plant is a spreading annual with procumbent branches and is to all appearances very much like mung or green gram. The stems and branches possess fairly long brown hairs, because of which it is known as the '**wooly pyool**' in some parts of the world. The leaves are large, trifoliate and hairy, generally with purplish tinge. The leaflets are broad, ovate and entire. The inflorescence consists of a cluster of five to six flowers at the top of a long stout hairy peduncle. The corolla is pale or lemon yellow. The plants are self-pollinated. The pods are about two or more inches in length. They are cylindrical and more or less curved. The pods are generally covered with fairly long hairs, longer than those of mung. The seeds are oblong in shape and are either black or green in colour.

Ecological Factors

Climate–It is grown as a rainfed crop in the warm plains and cool hills up to an altitude of 6000 feet. Since it requires a fairly warm growing season, it is usually grown in this country as a **kharif** crop. The black gram produced in the hills is better in cooking qualities.

Soil–The plant grows well on heavy soils. It prefers water-retentive loamy soils and does better on both black soils and brown alluviums.

Cultivation

The preparation of the land for sowing is similar to that of mung crop. It consists of two or three ploughings at the commencement of the rains. The seeds are then sown broadcast at the beginning of the rains. Harvesting is done in a manner similar to that of the mung crop.

Uses of Urd

1. It is used as a common pulse, since it is highly nutritious.
2. It is used in preparing 'papars', 'bari' (spiced balls of grounded dal), and delicious 'kachoris'.
3. It is medicinal. It is used in rheumatism and paralysis.

MOTH BEAN

(Mat bean; Vern. Moth)

(Phaseolus aconitifolius)

Moth Bean Growing Regions

The moth bean is indigenous to **India** probably recently domesticated. The bean is now grown in other parts of the world such as **China** and **United**

States of America. It is widely cultivated in India where it is grown in all parts of the country. Rajasthan accounts for four-fifth of the area and three-fourths of the annual production. Uttar Pradesh, Punjab and Maharashtra are the other important urd-producing tracts.

Botanical Description

The plant belongs to the family ***Fabaceae***. It is a herbaceous, creeping annual which spreads on the soil forming a mat like covering. Its mat-like character has given it the name of mat-bean in the United States of America. On account of its mat-like spreading habit, it is highly useful against wind erosion in sandy areas. The stem grows to a height of about a foot only with a spread of about two to three feet around the base. The branches are numerous and viny; the lower ones lie flat on the ground. The stems as well as leaves are covered with numerous hairs. The leaves are made up of three leaflets, each leaflet being divided into 2 to 5 very narrow segments. The leaves are similar in appearance to those of aconite plant; a poisonous weed. Hence it has been called **aconite-leaved bean**.

The flowers are in a cluster with a long peduncle, so that they appear above the leaves. But the individual flower pedicels are very short. Self-pollination is common, but cross-pollination may also occur. The fruit is a pod 1 to 2 inches, long, and the seeds are more or less cylindrical with rounded ends.

Ecological Factors

The crop is usually grown during the **kharif** season. It prefers well-distributed moderate rainfall, heavy rain being very detrimental to it. Moth is very resistant to drought. It is largely cultivated on well drained poor soils ranging from very light alluviums to heavy loams.

Cultivation and Harvesting

Its cultivation is more or less similar to that of green gram and black gram. Sowing starts with monsoon rains in June or July. The seeds are sown by broadcast method.

If the crop is grown for fodder, it is cut when the pods have been formed, but if it is grown for seed, the crop is harvested when the pods are ripe. The crop is cut with sickles is removed to the threshing floor and is threshed and cleaned in the usual way.

Uses of Moth

1. Unripe pods are consumed as vegetable.
2. The grain is used as pulse in the main producing regions.
3. Green as well as dry plants make good fodder.
4. Occasionally it is used as a green manure.

A brief account of some of the high-yielding varieties of pulses is given below:

Variety	Description	Areas of Adaptability
ARHAR		
Ageti	Erect and dwarf type maturing in about 150 days. Can be followed by wheat in rotation. Yield range 20-25 quintals/ha. Escapes frost, due to its maturity before December.	Throughout India.
Sharda	Bushy, semi spreading type, taking about 160 days to mature. Wheat can be sown in rotation after this crop. Yield range 20-25 quintals/ha. Escapes frost.	M.P, Tamil Nadu, Gujarat, U.P., W. Bengal, Maharashtra, A.P., Haryana, Mysore, Rajasthan and Bihar.
Mukta	Medium tall spreading type, Matures in 170-180 days. Resistant to Arhar wilt. Yield range 20-25 quintals/ha.	M.P, Gujarat,U.P., Tamil Nadu, West Bengal, Maharashtra, A. P. and Bihar.
MUNG		
Pusa Baisakhi	A short duration type, maturing in 65-70 days. Suitable for sowing after harvest of wheat, can be harvested before sowing of Kharif crops. Yield range 6-10 quintals/ ha.	Suitable for summer sowing all over India.
Mung P.S. -7	Early maturing having bold seeds. Maturing synchronously in 70 days. Yield 10-15 quintals/ ha.	Punjab, Delhi, U.P., Haryana, A.P, Bihar and Tamil Nadu.
LENTIL		
Pusa-1	Spreading type; having medium bold seeds, maturing 110-115 days. Yield 15-20 quintals/ha.	Haryana, Delhi, U.P, Bihar, A.P. and Orissa.
Pusa-6	Erect and compact type, having medium size seeds, maturity 120 days. Yield 15-20 quintals/ha.	Bihar, U.P., Delhi, Haryana and Punjab.
PS-16	Matures in 60-65 days; grains small and shining green; suitable for cultivation both in summer and kharif seasons Average yield 10 quintals/ha.	Haryana, Punjab, Delhi, U.P., Maharashtra and Gujarat.
URD		
Pusa-1	Early maturing type, maturing in 80 days. Yield 12-15 quintals/ha. Resistant to yellow mosaic virus.	Haryana, Rajasthan, Delhi, U.P., Bihar, Gujarat and M.P.

(*Contd.*)

Variety	Description	Areas of Adaptability
PEA		
Hans	Early maturing ; grains white and round ; suitable for multiple cropping and timely as well as late sowing conditions. Average yield 12.7 quintals/ha.	Pea-growing regions of the country.
P-410	Matures in 115-125 days; grains light green and medium bold; tolerant to powdery mildew and leaf minor Average yield 16 quintals/ ha.	Pea-growing regions of the country.
MASOOR		
Pusa-4	Matures in 130-140 days; suitable for timely sown conditions. Average yield 14 quintals/ha.	U.P., Haryana, Bihar and W.B.
Pusa-6	Matures in 130-135 days ; suitable for timely sown conditions. Average yield 15 quintals/ha.	Delhi, Haryana, Punjab, U.P., Bihar and W.B.

SOYBEAN
(Japan pea; Vern. Bhatt)

(Glycine max)

Origin and History

Soybean is the most important legume crop in the world. The soybean is one of the oldest crops grown by man. ***Glycine max*** is believed to have arisen from **G. *ussuriensis*** which is found in **South-eastern Asia**. It might have arisen from **G. *soja*,** a wild species growing in eastern China or both these species might have had a common ancestry. **G. *soja*** grows in Japan, Russia, Korea and Taiwan. Soybean has been mentioned in a Chinese book *Pen Ts'ao Kong Mu* by the Emperor Sheng-Nung as early as 2838 B.C. It was then believed to have been planted by the God of Agriculture, *Hou Tsi.* It is believed to have been domesticated in the eleventh century B.C. in the Kaoling region of Northeastern China. Due to large scale production of soybeans in Manchuria it was known as "field of beans". Soybean cultivation spread to Japan, Southeastern China, Taiwan, Vietnam, Cambodia, Laos, Thailand, India and Netherlands.

It was introduced in U.S.A. in 1804. It has since then become one of the chief crops of that country. The soybean is believed to have spread from China to eastern and southern Europe through Russia.

Soybean Growing Regions

WORLD. There are about 1000 varieties of soybean growing in eastern Asia alone. It accounts for one-quarter of the total agricultural land and half of all farm-income in Manchuria. The other leading soybean producers are Korea, Japan, China, Indonesia, U.S.A., Mexico, Canada, U.S.S.R., Rumania and Paraguay. Soybean exceeds an annual world production of over 40 million tons and accounts for about 200 thousand tons of oil. U.S.A. (74%), China (12%) and Brazil (8%) are the leading producers of soybean. U.S.A. has 60 million hectares under soybean cultivation. U.S.A. produces more than 32 million tons of soybean every year. It has become the most important cash crop in the United States, overtaking wheat and maize.

INDIA. Soybean is mostly cultivated in hilly areas of U.P., Punjab, Assam, Madhya Pradesh, Maharashtra, Kashmir etc. In India Soybean is grown in 8.25 MH and the yield is 8.67 MT.

Today soybean exists in thousands of varieties which are adapted to varying uses and differing conditions of the environment. They also differ in their growth habits, oil content of the seed, different maturation periods etc.

Ecological Factors

Soybean can be cultivated in a wide range of climatic conditions. It is responsive to the photoperiod. It requires short photoperiods for flower formation. Photoperiodic sensitivity results in restriction of a cultivar to a narrow belt of latitudes. Soybeans have been classified into 00-VIII groups on the basis of their day length requirements. The varieties of groups 00, 0 and I are adapted to longer days of North America and South Canada where the daylength is of more than 16 hours. On the other hand, varieties of the south flower under day-length of less than 10 hours. The plant can be cultivated upto a height of 2,100 m.

The development of such a large number of varieties of the plant is due to its long cultivation history and the very keen interest shown in the crop by the advanced countries of the world. The crop is widely grown in the tropics and the north temperate zone as a pulse. It is a very rich source of protein for a very large population of Asia. The varieties which are grown for oil have seeds with 16 to 19 per cent of fat content.

Botanical Description

Soybeans belong to the family **Fabaceae**. Soybeans are small erect, sub-erect, bushy or prostrate annuals. They grow usually to a height of 3 ft. The height may vary between $1\frac{1}{2}$ to 6 ft.

The leaves are trifoliate, the leaflets being fairly large (Fig. 4.6A). Some varieties shed their leaves before maturity, whereas others retain them up to the end. Small white or purple flowers are borne in clusters of few on short axillary racemes. The flowers are generally self-pollinated, but cross-pollination

A B C

Fig. 4.6. A. Soybean plant; B. Branch with fruits; C. Beans.

may take place under field conditions. The pods are 1 to $3\frac{1}{2}$ inches in length. Pods are hairy and are dark-brown in colour (Fig 4.6B). They bear two to four seeds (Fig. 4.6C). The seeds vary considerably in shape, size and colour.

The chemical composition of seed is highly variable. It has protein 40%, oil 20%, and carbohydrates 17%. Seeds in the lower part of the plant has higher percentage of these contents.

Cultivation

The seed is sown in June or July on well prepared seed beds. The seeds may either be sown in lines $1\frac{1}{2}$ or 2 feet apart by means of drills or corn planters. They may also be sown broadcast. To ensure availability of green fodder in October and November, the crop should be irrigated at least once after the cessation of monsoon rains. Seed germination is favoured by 375-500 mm of rainfall and temperature of 18-25°C. Excess of moisture checks aeration of soil and thus inhibits germination of the seed. Germination is epigeal. They can grow on all types of soils but prefer loams or clay loams for their proper development. The best results are obtained on fertile loam and sandy loam soils. Roots produce nodules to fix atmospheric nitrogen. Flowering occurs when the plant is 28—35 days old. Fruiting occurs within 8-20 days of flowering. All soybeans destined for commerce must have yellow seed coats and cotyledons. The plant matures in 75 to 150 days.

Harvesting

If grown for fodder the crop is cut when green and is fed to the cattle. When grown for green manure it is ploughed under when plants begin to bloom. But when grown for seeds, the crop is harvested before the plants are completely dry as most varieties shatter very badly when left too long in the field. Intensive breeding of superior varieties and use of the combine for harvesting have largely overcome this difficulty. After cutting they are brought to a threshing floor and threshed in the same way as that of the other pulses.

Uses of Soybean

The soybean is largely valued for its high protein (30-45%) and oil (20%) content. It is rich in calcium, iron and vitamins. Some of the important uses of soybean are as follows:

1. Soybean is the most important legume in the eastern Asia where it supplements the starchy rice in the daily diet. It is consumed fresh, fermented or dried.
2. The seeds are highly nutritious and constitute the richest vegetable food of the world.
3. The soybean flour with a low carbohydrate and high protein content is an excellent food for diabetics.
4. The flour has become a very important constituent of prepared food mixes in the breakfast food and bread.
5. Cooked soybeans mixed with roasted wheat flour, salt and a ferment give rise to a very tasty soybean sauce.
6. The seeds give soybean milk which is used in cooking. The milk is useful for the children and the invalids.
7. It is used in the form of liquid, powder and curd.
8. Immature green beans and their sprouts *e.g.,* soybeans sprouts (formed by germinating seeds) are taken as food.
9. It is also used as salad.
10. In the recent times soybean oil has assumed great importance as a cooking medium.
11. The oil consists of linolic and oleic components. It is used for many industrial purposes, *e.g.,* in the manufacture of soaps, glycerine, linoleum,

paints, putty, waterproofing materials, lubricants, adhesives, plastics, printing inks, insecticides, synthetic textiles and in leather dressing.

12. It is a good source of meal-valuable stock feed.
13. It is used for fertilizing sugarcane and rice fields.

LABLAB

(Bonavist Bean, Lubia; Vern. Sem)

(Dolichos lablab)

Origin and History

Lablab is native to India where it occurs wild. It has been cultivated since ancient times. It is cultivated throughout India for vegetable and as a pulse crop. It is extensively cultivated in other parts of **Asia** and **Africa**. In Sudan, it is also cultivated for forage. It is now an important legume in many tropical countries.

Botanical Description

The plant has about 50 to 60 species of which only 3 or 4 species are cultivated. The plant belongs to the family **Fabaceae**. It is a large trailing long-stemmed perennial plant. In cultivation it is often treated as annual. The woody twining stem bears trifoliate leaves alternately.

The flowers are white or purple in colour. The pods are flat broad structures, half an inch or more wide and two or more inches long. The shape and size varies in different varieties. They are sharpy beaked. The pods contain several large, oblong, glutenned seeds. They may be white, black, reddish, brown or mottled in colour with a prominent white aril.

Uses of Lablab

1. The pods and seeds are edible. They are cooked as vegetables.
2. The plant is grown as a cover crop. It is used for hay and forage.

GROUNDNUTS

(Peanuts, China badam, Vern. Moongphali)

(Arachis hypogaea)

Origin and History

Groundnut is a highly nutritious legume crop. The seeds have protein (25-28%) and oil (43-55%). The high oil content of the seed has made groundnut an important oil-yielding crop

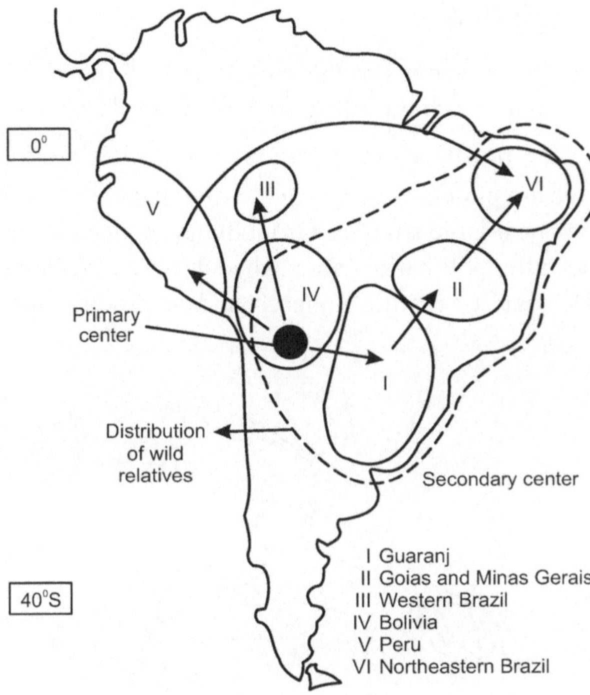

Fig. 4.7. Centers of origin and distribution of ***A. hypogaea***.

The groundnuts were first cultivated in **Brazil** (South America) in prehistoric times. Five kinds of groundnuts grow wild in Brazil. There are six secondary or tertiary centres of origin and diversity of ***Arachis hypogaea*** (Fig. 4.7). From Brazil, its native place, the groundnuts were introduced into **Peru**, where they have been found in graves 2,400 to 3,000 years old. Traders carried them to Mexico and West Indies. Portuguese sailors are believed to have carried the groundnuts from Brazil to Europe and Africa. The seeds were introduced to East Indies and the Philippines in 1500's and early 1800's. The crop was introduced in U.S.A. by middle of last century. Groundnuts came to India about the year 1800's. Since then it has spread very rapidly specially in the black cotton soil areas where it is being used as a rotation crop for cotton.

Groundnut Growing Regions

World–The chief groundnut growing regions are India, China, Indonesia, East and West Africa, U.S.A., Polland and France.

India–India occupies the top position in the world with regard to acreage and production of groundnuts which are 6–7 million hectares and 5–7 million tons respectively. The States of Gujrat, Maharashtra, Andhra, Karnataka and Tamil Nadu account for about 90 per cent of groundnut-producing area.

Botanical Description

Arachia hypogaea belongs to the pulse family **Fabaceae**. The plant is a bushy annual with the branches trailing on the ground. The central stem is more or less upright. The prostrate or the sub-erect branches are $1\frac{1}{2}$ to 2 feet in length. The leaves are pinnately compound with quadrifoliate leaflets arranged in two pairs. They bear large stipules. Groundnuts are peculiar in that the small air-borne flowers, after pollination, are pushed down by on elongating pedicels into the soil. The ovaries mature underground to produce the characteristic shelled peanut (Fig. 4.8).

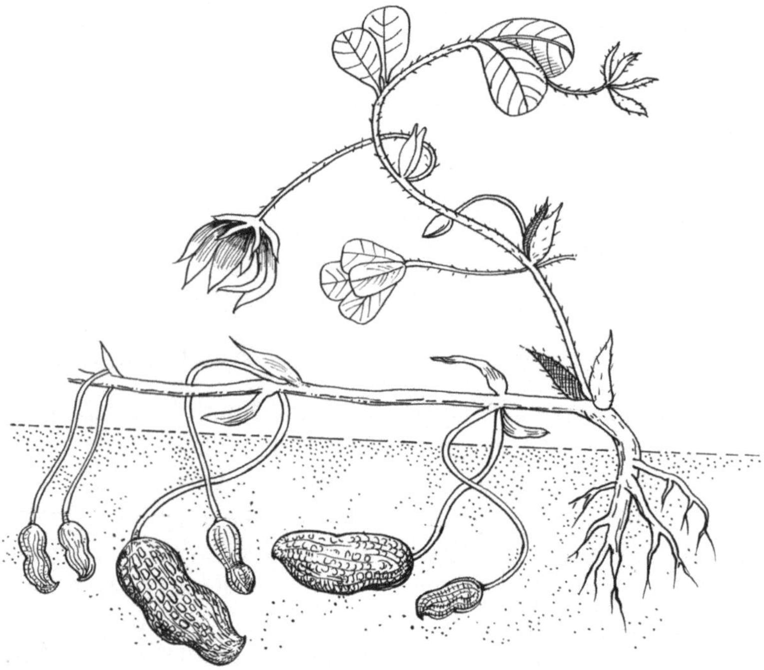

Fig. 4.8. A groundnut plant showing characteristic underground pods.

It is very essential that the ovary be able to push itself into the soil otherwise fruit formation does not take place. Ordinarily groundnut flowers are self-pollinated, but cross-pollination may take place due to the visitation of the flowers by insects.

The fruit is oblong, more or less constricted pod of pale straw colour. It varies in size according to the variety. The pod contains one to several ovoid seeds. The shell of pod which contains the seeds is morphologically the pericarp and the thin skin which covers the seed or the nut is the seed coat or testa. The testa varies in colour in different varieties. It may be usually brick-red or rust brown. The nut is composed of two cotyledons. It is these structures which contain the oil and other food materials.

Groundnut is rich in oil (45%) and protein (25%). Oleic acid (56%), linolenic acid (25%) and palmitic acid (6–12%) are the principal fatty acids. **Arachin** and **conarachin** are the principal proteins present in groundnut. Groundnut has maximum caloric value.

Ecological Factors

Climate–The groundnut is essentially a tropical plant and, therefore, requires a long warm growing season. While the plant requires a certain amount of moisture (moderate rainfall) an excess of water is detrimental to its growth. It can be grown successfully in areas having a rainfall as low as 20 inches and as high as 50 inches. The plant can bear neither frost nor long and severe drought or water stagnation. During its ripening period it is essential that there should be about a month of warm dry weather. It seems that the plant grows best where the mean temperature is from 70° to 80°F. Lower temperatures are not suitable for its development, and frost is very detrimental to the groundnut crop.

Soil–The crop grows best on light sandy soils so that the ripening fruit can be buried. Sandy loam, loam and well drained black soils are good for the crop. Heavy and stiff clay soils are unsuited to it. It thrives best on those soils which are well manured and are rich in organic matter.

Cultivation

Groundnut is raised mostly as a rainfed **kharif** crop, being sown from April-May to June-July depending upon the receipt of monsoon rains. An irrigated crop is grown to a limited extent in South India between January-March and May-July.

The land should be thoroughly ploughed. The first ploughing may preferably be done immediately after the **rabi** harvest. The land should be worked until it is fairly well pulverized. The seeds are generally sown after they have been shelled, but sometimes unshelled pods containing only one seed per pod are selected for sowing. In some case such one-seeded pods are cracked and sown without removing the shell. It is, however, always advisable to sow shelled nuts as this allows one to judge their quality.

The method of planting groundnuts varies in different regions. In certain areas the seed is drilled while in some other areas the seed is dropped in the furrow, opened by the plough. The seeds when dropped in the furrow are covered by means of a planker. The covering of the seed is very necessary, as otherwise they may be picked up by birds or other animals. When planted in lines the distance between lines is about one foot, and the distance from plant to plant in the lines is from 4 to 9 inches.

Groundnuts, being a leguminous crop, usually do not require any manure. Experiments in certain regions show that an application of sulphate and potash is profitable for groundnuts. The application of lime has also proved beneficial to the crop.

Bhabha Atomic Research Center (BARC) has developed giant groundnut varieties such as gamma radiation mutants TG-1, TG-3 and TG-18. TG-1 has a kernel size double the size of the popular Spanish-improved type. TG-18 has a kernel size which is even bigger than TG-1.

Harvesting

Groundnuts should be harvested when fully ripe. If harvested before this stage there is a high percentage of fatty acid which has not been converted into oil yet. It has been found that the harvesting of groundnuts even one week before the nuts are fully ripe decreases the oil contents by about five per cent.

The maturity of the crop is usually indicated by the yellowing and shedding of the lower leaves. Ripe pods may be distinguished by their darker colour and comparative dryness. However, all the pods in the plant do not mature at the same time. Nevertheless the crop is usually harvested at one time. The bunch variety is harvested by pulling out the plants while the spreading variety is harvested after loosening the soil by digging, ploughing or by light irrigation. The pods are stripped from the vines immediately after harvest or after spreading the plants for few days for drying. The pods are dried in the sun before storage.

Uses of Groundnuts

1. Its economic importance has increased tremendously as the demand for vegetable oil has increased. It contributes about one-sixth of the world's supply of vegetable oils. After soybean it is the largest source of vegetable oils. Most oil extraction takes place in Europe and India. It is high in oleic, modest in linolic, and low in certain other fatty acids. This vegetable oil is extensively used for cooking purposesand also for the manufacture of hydrogenated Vanaspati (Ghee), soap and toilet requistes.
2. Groundnuts are a very nutritious food since one lb. yields 2700 cal., whereas 1 lb of beef furnishes only 900 cal.
3. Kernels are also eaten raw, roasted or sweetened. They are rich in proteins and vitamins A and B and some members of the B_2 group.
4. The oil cake resulting from expression of the oil is one of the best known stock feeds, since it has a high protein contents, higher than that of any

other oil cake. Many pastures in South America are covered with peanut vines instead of grass.

5. Peanut flour, peanut milk and peanut butter are prepared from the seeds of groundnut.
6. Groundnut oil is used for cooking sardines before packing them in olive oil.
7. Inferior groundnut oil is used in the manufacture of lubricants, soaps, etc.
8. Groundnut oil is a laxative.
9. The crop is also of value as a rotation crop. Being a legume with root nodules, it can enrich the soil with nitrogen and thereby improves soil fertility.
10. Peanut butter is a very nutritious preparation.
11. The groundnut protein is used in the manufacture of a synthetic fiber called ardil.

Diseases

The plant is subjected to a number of fungal diseases. One of the important diseases is the Tikka disease caused by ***Cercospora***.

QUESTIONS

1. What type of climatic conditions favour the cultivations of groundnut?
2. Describe the origin, distributions and cultivation of soybean.
3. Describe the botany and method of cultivation of any three pulse crops.
4. Why are pulses so important in our food? Explain
5. Give the botanical names of any five legumes and discuss the ecology of any one of them.
6. The Govt. of India has given top priority for the improvement of pulses in India in the coming years. What, in your view, are the objectives of the programme?

5

Food Plants (Contd.)

NUTS

Though a variety of plant structures like seeds, legumes or dry drupes are also sometimes called nuts, botanically a true nut is a one seeded, one chambered dry fruit with a hard pericarp. Unlike the cereals and the legumes the nuts are borne by woody perennials. Some plants can grow wild whereas others are cultivated.

While cereals and legumes are rich in carbohydrates and proteins respectively the nuts contain plenty of fats. Though nuts are a valuable food material they are seldom staples in the diet. Besides fats, the nuts are rich in protein and mineral elements and may also contain starch, and sometimes sugar. Nuts, therefore, constitute a well-balanced diet.

Nuts can be handled roughly and can be stored very well owing to their very low water content. Nuts are very delicious to eat and have, therefore, assumed great importance in the recent past. An interesting point about their cultivation is that unlike the cereals and the legumes they can be grown on non-agricultural land.

The nuts are arranged under three groups irrespective of their morphological nature:

(i) **Nuts with a high fat content:** Brazil Nut, Cashew Nut, Coconut, Hazelnut, Hickory Nuts, Macadamia Nuts, Pecan, Pili Nut, Pine Nuts and Walnuts.

(ii) **Nuts with a high protein content:** Almonds, Beechnuts and Pistachio Nuts.

(iii) **Nuts with a high carbohydrate content:** Acorns and Chestnuts.

A description of cashew nut is given below whereas coconut is described in chapter on oil yielding plants.

CASHEW NUT
(Vern. Kaju)
(Anacardium occidentale)

History and Distribution

Cashew nut is a native of **Brazil** from where it spread very widely in the tropical countries from West to East by the Portuguese explorers. The chief cashew nuts-growing countries are Mexico, Peru, Brazil, West Indies, southern Florida (U.S.A.), the countries around the Mediterranean, Mozambique, India, Malaysia, Indonesia, etc. In India, it is chiefly grown in Kerala, Karnataka and Goa. India is the main commercial producer of cashew kernels. The main exporting countries are India, Mozambique and Tanzania and the largest importer is U.S.

Botanical Description

The cashew nut belongs to the family **Anacardiaceae**. The plant is a small tree. The plant is vegetatively propagated by several different methods such as layering, inarching, side grafting, patch and shield budding. The tree bears small unattractive flowers and pear shaped 'fruits' called cashew apple. The fruit is thin-skinned yellow or reddish coloured and is juicy. It is in fact a false fruit made up of swollen peduncle and disc. The true fruit which is the cashew 'nut' is borne at the distal end of the 'apple' in the form of a small curved or kidney shaped structure (Fig. 5.1). The cashews are roasted and the seed coats are removed before the nuts are sent to the market. India exports a large amount of cashew nuts to U.S.A. and Russia.

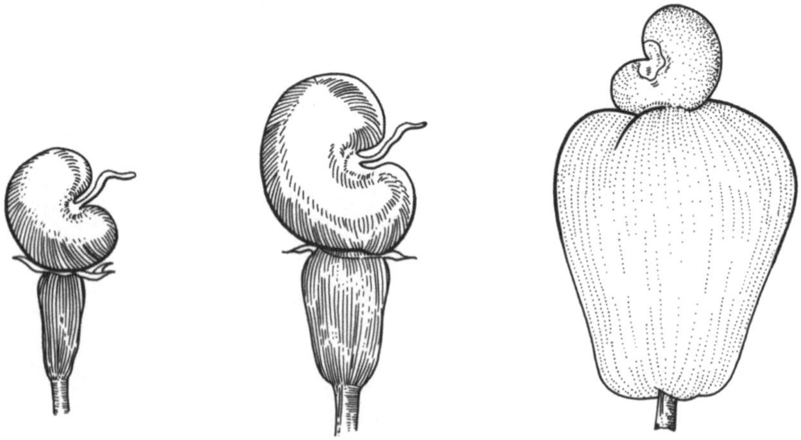

Fig. 5.1. A cashew nut showing formation of the swollen peduncle apple and the reniform nut.

Ecological Factors

The cashew nut tree has commonly been regarded as one of those hardy or tough trees, which are able to grow under conditions too unfavourable for other trees. The tree has become naturalized in many countries and grows readily often in sandy and stony areas unsuited for other crops. Furthermore, near the sea it withstands salt-ladden winds better than other trees.

Cultivation and Harvesting

The cashew nut tree is usually raised from seed since the seed germinates easily and the seedlings grow even under apparently adverse edaphic and environmental conditions. In the recent years plants have also been propagated by different vegetative methods.

Harvesting is done by hand. Formerly the nuts were roasted in open pans before shelling, but this method was superseded by perforated cylindrical rotary roasters, which give a more uniform produce. More recently oil bath processes have been employed which give quicker and more uniform roasting. The recovery of the shell oil, which has several industrial uses, is also higher in this techrique.

The kernels vary in size (there are about half a dozen grades). At the Cashew Research Station, Ullal, Mangalore it has been shown to range from 48 to 212 in a pound (average 86).

Uses of Cashew Nut

1. The 'apple' is fermented into Kaju wine.
2. The seeds yield a very nutritious oil.
3. The ripe fruits, the nuts are bland sweet in taste. They are eaten raw or after they are roasted with salt. They are softer and more easily chewed than many other nuts. The development of the vacuum pack in hermetically sealed tins has established the cashew trade. Previously, the kernels were damaged by insects during transportation to overseas or even inland markets.
4. The light close-grained wood, the leaves and the bark are useful.
5. The shell oil has many industrial uses *viz,* in water proofing and preservatives, etc.

Diseases

A recent survey of insect pests of the cashew tree in India has revealed the fact that it may be attacked by no less than 40 different insects, some of them inflicting heavy damage.

Some other Nuts

1. **Walnut** *(Juglans regia ;* Vern. Akhrot).
2. **Almond** *(Prunus amygdalus;* Vern. Badam).
3. **Bitter almond** *(P. amygdalus* ;' Var. *amara).*
4. **Pistachio Nut** *(Pistcaia vera.* Ver. Pista).
5. **Chestnut** *(Castnea dentata).*
6. **Hazelnut** (*Corylus americana).*
7. **Beechnut** *(Fagus grandiflora).*

QUESTIONS

1. Why are nuts considered to be highly nutritious?
2. Discuss the botany, distribution and cultivation of cashew nut.
3. Name five nut yielding plants. Give their botanical names and uses.

6

Food Plants (Contd.)

VEGETABLES

Vegetables constitute an important part of the daily diet. There are a variety of plants whose edible parts are eaten raw or are cooked for preparing delicious dishes. The vegetables include roots, stems, leaves, flowers and even fruits. Though they contain plenty of reserve food, their food value is not much because of the presence of large amount of water (70 to 95 per cent). Starch is the main reserve food in vegetables. The vegetables contain mineral salts and vitamins. Some of the important vegetables are described below under the following heads:

(i) Underground vegetables.
(ii) Herbage vegetables.
(iii) Fruit vegetables.

I. UNDERGROUND VEGETABLES

There are many plants which have their underground structures modified into storage organs. These underground fleshy parts can perennate under unfavourable conditions. Morphologically such structures are modifications of roots or stems.

Root Vegetables

CARROTS

(Vern. Gajar)

(Daucus carota)

Origin and History

Carrots are known to have been cultivated for over 2,000 years. Nothing is known about their origin. One reason for this is the fact that they have at least twentyfive close relatives. Several of these are believed to have interbred, probably in **Central Asia** (Afghanistan etc). Traders took seeds westward to

Africa and Europe. Carrots spread eastward too, but not so rapidly. The carrots were known to the Greeks and Romans in the early Christian era. The plant was introduced into England by 1600 A.D. Ladies at the court of King James I put carrot leaves on their hats and in their hairs. Both explorers and early settlers brought carrots to Virginia (North America) in 1609. Soon it spread over to the whole of America. Carrots are now seen growing practically in the entire world, wild, naturalized or cultivated.

Botanical Description

The plant belongs to the family **Umbelliferae**. Carrots like beets are biennials but are ready for harvesting in one year. The swollen tap root possess wide cortex and phloem layers which store the reserve food. The root has typical conical appearance (Fig. 6.1). The roots of Japanese variety are 3 ft. long. The plant bears characteristic pinnately compound leaves. It bears an erect umbelbearing stem in the second year to produce flowers and seeds. Although most carrots are dark yellow or orange, some have other colours. It normally varies from reddish purple to white. Carrots vary in shape, size, colour and quality depending upon the soil conditions.

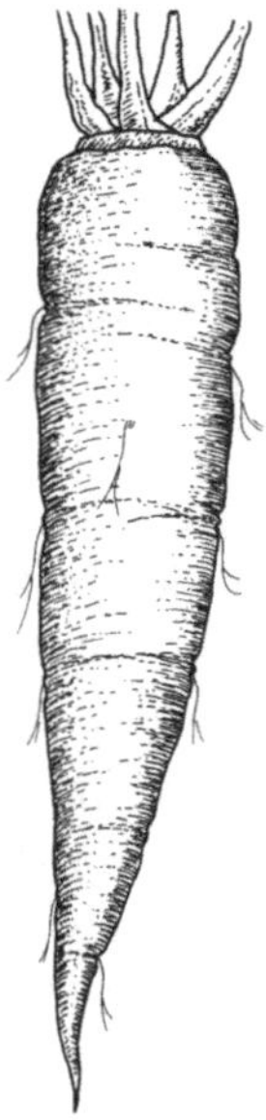

Fig 6.1. A carrot.

Cultivation

Carrots grow best in deep sandy loam soils. They are planted in rows, thinned, pulled by hand, and stored like other root crops.

Uses of Carrots

1. In earlier times carrots were used for medicine. Greek books about 1,900 years old, mentioned that carrots were a 'stomach tonic'. Both Greeks and Romans prepared medicine from carrot seeds and used it to treat snake bite. Roots were made into poultices for cancers.
2. The carrots have been in use as stock-feed.
3. Carrots are eaten raw or cooked as vegetable because of their abundant carotene.
4. 'Gajar ka Halwa', a sweet preparation is relished very much by north Indians.
5. Raw carrots are eaten as salad.
6. 'Gajar ka Kanji'is taken by Indians as an appetizer. It is prepared by pickling black carrots in salt and rye.

RADISH
(Vern. Mooli)
(Raphanas sativus)

Origin and History

According to one view radish has originated somewhere on the eastern side of the **Mediterranean**. According to another view radishes are believed to have grown as weeds is **ancient China** from where the cultivated species were taken westward across Asia to Egypt as much as 4,000 years ago. Early Egyptians are reported to have eaten radish roots as vegetables and extracted oil from their seeds. A temple at Karnak on the river Nile bears images of radishes on its walls. The ancient Greeks became so fond of radishes that they offered them to their god Apollo in golden dishes. Radishes were introduced in England before 1550 and by 1630 the colonisers took them to America. Radishes of various sizes and colours are now grown in different parts of the world. In India radishes are now grown very extensively. Maximum radishes are grown in Japan and China.

Botanical Description

Radishes are annual or biennial plants. They belong to the family **Brassicaceae**. The stem is reduced and is in the form of a disc. It bears a rosette of small leaves above and a fleshy tap root underground (Fig. 6.2). One old book of sixteenth century reports a winter radish that weighed 100 pounds.

Radish is perhaps the most quickly and easily grown vegetable of the kitchen garden. The edible roots may be white, red or black. They are fusiform in shape. The size of radishes varies considerably.

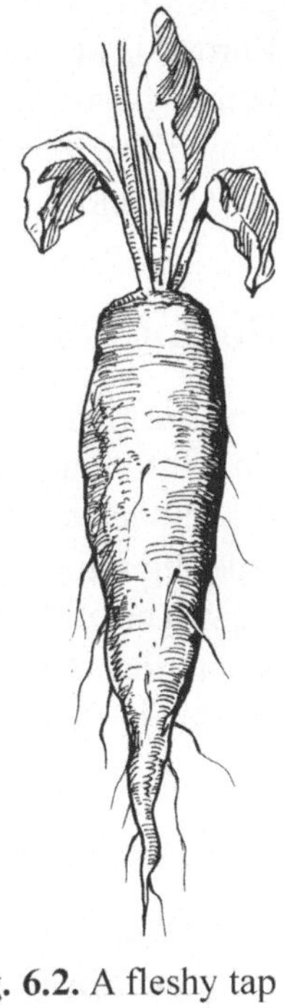

Fig. 6.2. A fleshy tap root of radish.

Cultivation and Harvesting

Radishes are grown as winter, spring, late summer and autumn varieties in different countries. Most Americans do not like winter varieties. Europeans keep winter radishes in cool, dark cellars and eat them when the weather is cold. People in cooler parts of Asia eat winter radishes. The Japanese grow radishes 2 to 3 feet long and 4 inches thick. A radish grown in India has slender seed pods about 4 feet in length. Radish is harvested by hand, cleaned and washed before it is sent to the market.

Uses of Radish

1. In Japan and China, the two chief radish-growing countries, the crop is usually pickled in salt water (brine) to become a staple in the diet.
2. The radishes are eaten raw in India and in many other countries.
3. They are also cooked as a fresh vegetable.
4. Radishes are used as an appetizer.
5. The young, tender leaves are used as salad or are cooked as vegetable.
6. 'Mungre', the fruits of radishes are also cooked as vegetables.
7. Radishes are also grown as a stock-feed.

TURNIPS
(Vern. Shaljam)
(Brassica rapa)

Origin and History

Turnips have been in cultivation for over 4,000 years. They are believed to be native of **Eastern Europe** and **Central Asia** (**Siberia**) where they still grow in a wild state. The time when turnips were carried to north western Europe is not known, but they were common in France, Holland and Belgium before 1400. Turnip seed was carried to Canada in 1540 by a French explorer named **Jacques Cartier**. Turnips were introduced in Mexico in 1586, in Virginia in 1609, and in New England in 1628. Turnips are now worldwide in distribution. It is a common vegetable in India.

Botanical Description

Turnips belong to the family **Brassicaceae**. They are annuals. The stem is extremely reduced. It bears leaves above and a large spherical, sharply tapering napiform root below (Fig. 6.3). The roots vary in size, shape and colour. Though some turnips are almost ball-shaped, most varieties are flattened, and their flesh is generally white. The leaves are thin and rough and are covered with prickly hairs.

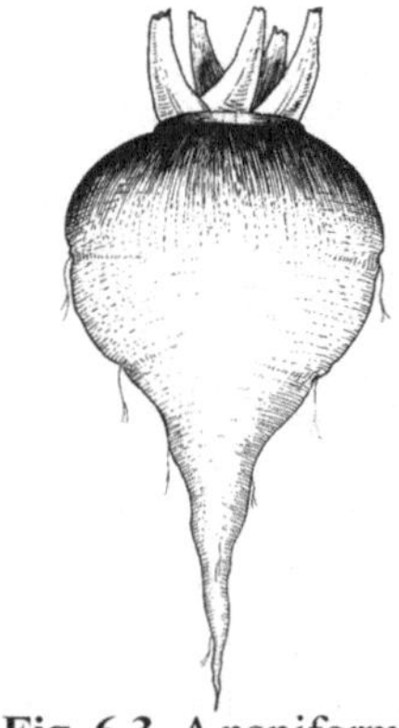

Fig. 6.3. A napiform root of turnip.

Cultivation and Harvesting

Turnips thrive best in cool weather. The plant is very easily cultivated and is known for its hardiness. It is sown broadcast or in drills. The seedlings are thinned to 6 inches apart. The crop is harvested by hand. The turnips are cleaned of soil particles, and are washed before sending them to the market.

Uses of Turnips

1. Turnips are cooked as vegetables.
2. While the soft turnips are used by man, the coarser ones are fed to the stock.
3. Turnip leaves are used as salad and for forage and manuring.
4. Turnips are boiled and then pickled by adding salt, spices, etc.

BEETS
(Vern. Chukander)
(Beta vulgaris)

Origin and History

Beets were domesticated first as a leaf vegetable, then as a root crop, and finally as a source of sugar.

There are a number of varieties of beets in cultivation. The modern beets are believed to have originated in **southern Europe** along the seacoasts of the **Mediterranean region** from the wild beet *(B. maritima).* They were grown in Germany in about 1558. Beets were carried to England from Italy in 1576. French people are reported to be growing modern beets as early as 1631. Beets are now very widely cultivated. The chief beet-growing regions are Germany and other European countries, Russia, Asia Minor, North China, Japan and United States.

Botanical Description

The beets belong to a single species ***Beta vulgaris*** *of the* family **Chenopodiaceae**. The common garden beet is a biennial though as a garden vegetable it can be grown as an annual. The plant bears a large turnip like fleshy root. The stem is extremely reduced. It bears large petiolate leaves at the top. The major part of the flesh of the root is composed of xylem and phloem parenchyma. Seeds are produced, in clusters. There are four kinds of beets:

1. **Chard**–It is the oldest beet known (300 B.C.).
2. **Mangel wurzels**–It has evolved from Chard. They are grown very widely in Europe and Canada. They are very large, pinkish yellow beets with coarse roots and big leaves.
3. **Sugar beets**–They have arisen from Mangels. They are smaller in size. They are grown in Europe, U.S.A. and Asia, and are used for manufacturing sugar.
4. **Common beets**–A number of varieties of different shapes, sizes and colours are cultivated.

Cultivation and Harvesting

Beet particularly sugar beets have received considerable attention since the time of Napolean in the early 1800's. The beets are cultivated on almost all kinds of soils. Sugar beet is commercially cultivated in the **Temperate Zone**.

The beets are drill planted in rows in early spring. The seed clusters are broken so that the seedlings are not very dense. The seedlings are usually thinned out by hand or by mechanical machines equipped with a rotating propeller-like blade.

Mature crop is harvested by mid-autumn by hand or by mechanical means. A very sophisticated modern machine called the spiked wheel does the digging, topping, loading on trucks, and windrowing of tops of sugar beets. Sugar is extracted from harvested beets by diffusing and recovery processes.

Uses of Beets

1. Sugar beets account for one-third of the world's total production of sugar. Sugar beet industry is competing successfully with sugarcane industry in meeting world's increasing demand of sugar. An otherwise ordinary vegetable has become a sugar crop of worldwide importance due to efforts of plant breeders. The modern sugar beets have been bred for higher sugar content.
2. Thin roots of wild chard were once used for medicine.
3. Chard beats are popular as a pot herb.
4. The leaves are eaten for greens.
5. The beet root is cooked as vegetable.
6. The red beet is cut into slices to be eaten raw as salad.
7. The roots arc pickled, boiled or canned.
8. Mangels are fed to cattle dry or are used for silage.

SWEET POTATOES

(Vern. Shakkarkandi)

(Ipomoea batatas)

Origin and History

Sweet potatoes originated from a wild vine that still grows in **tropical America** (sometimes called **Central America**). Red Indians cultivated this plant about 2,000 years ago in the region lying between Mexico and the Isthmus of Panama. The slender vines and the small tubers are believed to have become larger in course of time, so that the present day tubers contain plenty of starch and sugar.

Sweet potatoes were carried by traders to Peru in the south, Mexico in the north, and Caribbean islands in the east. Columbus took sweet potatoes to Spain from Cuba. Later explorers also carried them to Europe. By 1650, i.e., only 57 years after Columbus brought them to Spain, red, purple, and whitish sweet potatoes were growing in most of the warm regions of Europe. While settlers from England carried them to U.S.A. by the end of sixteenth century, Spaniards carried them to the Philippine Islands soon after 1565. Soon it reached China, India, Indonesia and Japan. Sweet potatoes became so popular in Japan that it now accounts for half of the world's total production of sweet potatoes. It is the second most important crop of Japan next only to rice. Recently it has been introduced into Australia and Russia.

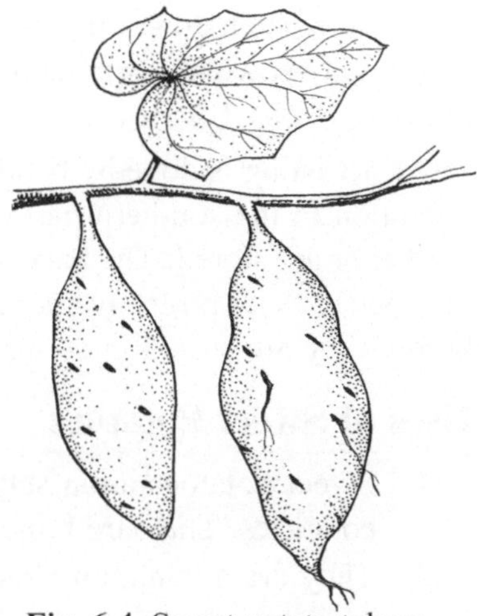

Fig. 6.4. Sweet potato tubers.

Botanical Description

Sweet potato belongs to the family **Convolvulaceae**. The plant is a twining, trailing vine. It has heart-shaped, frequently lobed leaves which are alternately arranged. They have blue or yellow flowers that resemble other morning glories. It is a perennial plant though it can be cultivated as an annual. In tropical countries, the plants live for several years (perennial) and grow very large. In cooler temperate regions, sweet potatoes seldom bloom and their vines are killed by frost in the fall. New plants are grown from cut pieces of vine, or from sprouts that grow on tubers kept in hot beds while the weather is cold. The plant produces adventitious roots, the tips of which swell up in the form of large tubers (Fig. 6.4). The sweet potatoes have usually pale-brown skins though grey, white, pink and purple skins are also quite common.

There is a starch-rich cortex within the outer epidermis or periderm of the sweet potato. The parenchyma cells constitute the flesh in a mature root. Sweet potatoes are very nutritious since they contain besides 70 per cent water. The tubers have starch = 18%, other carbohydrates = 8%, protein = 2%, minerals = 1%, (iron, calcium, and vitamins, etc.).

Cultivation and Harvesting

Sweet potatoes are cultivated in tropics of both hemisphere and to some extent in the temperate regions as well. They can grow very well in any kind of soil.

Sweet potatoes are planted by hand. Usually the pieces of roots or vine cuttings are sown at warm temperatures (70 to 75°F). In Japan planting is done between rows of barley.

Harvesting is done by hand. Mechanical harvesting has been brought into operation in the southern part of U.S.A. The potatoes are sent directly to the market or are stored. The harvested potatoes are kept warm by placing them in the open so that a protective periderm is formed. They are treated with fungicides before they are stored in cool places.

Uses of Sweet Potatoes

1. Sweet potatoes are a staple crop in southern U.S.A. and in many poor countries. They are boiled before eating.
2. They are a common vegetable.
3. They are used for preparing a cornmeal-like form in Japan.
4. Sweet potatoes are fermented to produce alcohol.
5. They also give starch and glucose syrup.
6. The tubers are a very good stock feed.
7. The green tops may be used in silage or for animal fodder.

CASSAVA, MANIOC, TAPIOCA, MANDIOCA, ETC.

(Manihot esculenta)

Origin and History

Cassava is second in importance to the sweet potato as a root crop in the tropics. Like potato, sweet potato, yams and aroids, tapioca is a starch-rich underground structure. It is a root vegetable like sweet potato and yams and constitutes staple food of tropical peoples in Latin America Africa and Southeast Asia. Nearly 500 million people of 4,000 million world population eat cassava. It is popularly called 'tapioca' in Asia. In other countries tapioca word is used for the manufactured products of the starchy roots of the plant. The crop is also known as 'manioc', 'mandioca' in some countries.

The crop occurs in a natural state only in the western hemisphere in the Americas between 33°N (USA) and 33°S (Argentina). Cassava is native to **America** and is believed to have originated in **eastern Brazil** in the second or third millenia B.C. It is assumed that the ancestors of cassava were one of the first food plants used by the men who migrated into Central and South America. Since the plant is not found in the wild state it is believed to have been domesticated in the remote past. The exact ancestry of cassava is not known. The species which show maximum resemblance to ***M. esculenta*** are the wild species. ***M. pringle***, ***M. aesculifolia*** and ***M. rubricaulis***. The first one has a low content of the poisonous cyanogenic glucosides and, therefore, it may be the possible ancestor of cassava. Hybridization and selection might be responsible for the variability in the cultivated forms.

Today cassava has spread all over the wet tropics except at the highest altitudes. It spread from South America in post Columbian times. By the middle of nineteenth century it become a popular food crop in the Old World. It has been widely grown as a famine reserve because of its tolerance of drought; locust attack and poor methods of cultivation. The crop arrived in India in 1800. It was introduced in Kerala about 150 years ago when the Maharaja of Travancore, Shri Vishakham Tirunal grasped the potentialities of cassava and propagated it in his state.

Distribution

It is widely grown in all the tropical countries of America, Africa and Asia. It has very easily adapted to new environments and today over 200 varieties of the plant are cultivated throughout the world.

Cassava is widely grown in Kerala and Tamil Nadu. It is only second to rice as a staple food in this state. Kerala produces about 5 million tons of cassava annually and accounts for 80 per cent of the total production of this crop in our country. In Tamil Nadu the area covered under the cassava cultivation has recently gone up to more than one lakh hectares from a mere forty thousand. Other states such as Karnataka, Andhra Pradesh and Maharashtra have also started extensive cultivation of cassava. The crop is also now grown in the hilly regions of Assam and hilly states of Northern India. It is a source of foreign exchange earnings. It has been successfully cultivated in the barren stream beds of seasonal rivers in the valleys of outer Himalayas (Dehradun valley, etc.).

Botanical Description

Cassava belongs to the family **Euphorbiaceae**. The plant is a parennial shrub. The plant usually attains a height of 6-9 ft. The stem is woody. It is glabrous with various degrees of branching. Some of the varieties are dwarf and have a height of not more than three or four feet. The branching may be all over the stem or may be restricted to the base of the stem giving a bunched appearance.

The stem shows variation in colour from silvery to reddish or dark brown. The plant bears large palmate compound leaves (Fig. 6.5). They are alternately arranged with long petioles. The lamina is deeply lobed to produce 3 to 7 leaflets. The morphology and colour of the leaves are highly variable. The roots are well developed, and the main roots develop large tubers. The tubers have a thin periderm, cortex, a small vascular cylinder and a very large starch-rich pith.

Fig. 6.5. Cassava plants.

Cassava plants rarely flower. The flowers are very small. They are unisexual and occur in terminal panicles. The calyx lobes are coloured and the corolla is absent. The male flowers have ten stamens in two whorls of five each. The gynoceium in the female flower is tricarpellary, trilocular with one ovule per locule. The fruit is a globular capsule with six longitudinal wings.

There are two kinds of cassava—bitter cassava and sweet cassava. Some varieties are poisonous. This is due to presence of the cyanogenic glucoside which breaks into the poisonous **prussic acid**. The "bitter" varieties have the glucoside distributed all over in the tuber while the "sweet" varieties have their storage parts relatively free of them. The bitter varieties are extremely poisonous for the cattle as well as the human beings. The poisonous substance can, however, be removed by cutting the tubers into pieces and boiling them for a long period. They are also squeezed to expel the prussic acid. The tubers may be grinded to a powder form and then pressed.

In the "sweet" varieties the rind is removed before boiling. The "bitter" varieties are more nutritious. It should be noted that the bitterness and sweetness of a variety is not a characteristic feature and is changeable under different climate and soil conditions. High-yielding varieties are H-2304 and H-1687. They give an yield of 34 to 46 tons per hectare-under rain-fed conditions.

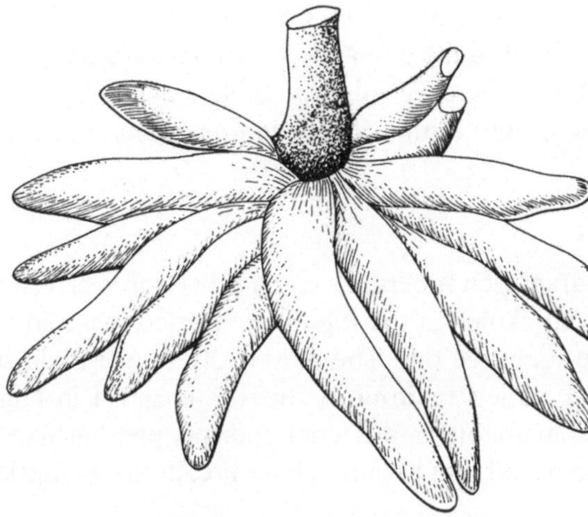

Fig. 6.6. Roots of cassava.

Cultivation

The plant is cultivated in hot climates. It grows very easily and does not require much labour. Stem cuttings of few inches in length are sown in holes dug at intervals of 1 metre. Different varieties take different time to mature. Some varieties produce mature tubers within six months to a year while others take 18 months to 2 years to mature. Secondary growth occurs in the adventitious roots to form tubers.. Tubers have a starch content of about 30 per cent. Older tubers are not edible due to lignification of the cells. The tubers are highly variable in size. They are very elongated and cylindrical structures attaining a length of two or three feet (Fig. 6.6). The tubers develop in a cluster around the base of the stem in well grown plants. The tubers may be surrounded by a mass of normal fibrous roots. The tubers are usually harvested after a gap of 8 months but large and better quality yields can be obtained after a period of 16 months or so. The left-over roots continue to produce new stems so that the plant remains perennial. **Central Tuber Crops Research Institute** (CTCRI) at Trivandrum has successfully introduced the **Mukhibhat System** of growing cassava. This system is named after an Indonesian farmer. This system involves making cassava grafts using ceara rubber as scion on hybrid varieties used as root stocks. The tubers of such grafts are much larger in size.

The Institute has also developed a technique by which 100 to 125 plants can be got from a single stem to reduce the waiting period for breeders by half.

Extraction of Starch

Cassava is an excellent source of tapioca starch. The tubers are first peeled and grated. The milky juice is removed. The starch-rich pulp is soaked in water for a few days. It is pressed into a mass and is then filtered to remove fibers and impurities. It is dried in the sun and is then heated gently in iron plates at controlled temperatures. Starch is slightly cooked and is hydrolyzed partially into sugar. It swells up into lumps. These lumps of starch are sold as tapioca '**pearls**'.

Uses of Cassava

1. The carbohydrate-rich tubers are eaten raw or are cooked as vegetables. In eastern South America the tubers are peeled, washed, shredded, and the juices are squeezed out. The pulp is dried over open fires to yield a flourlike meal, called **farinha** in Brazil and **garri** in Nigeria. It is the cheapest food available to the local poor people. Farinha may be eaten dry or may be baked into breads. These breads are as nutritious as wheat breads.
2. The juice of bitter varieties is build to produce '**cassareep**' which is used in preparing sauces and "West Indian pepper pot".
3. The juice is fermented into alcoholic beverages.
4. The tapioca starch is used in preparing syrups, alcohol, acetone, etc.
5. The starch is used as animal fodder.
6. The starch is used for sizing cloth and as a raw material for making mucilage.
7. The poisonous cassava contain poisonous hydrocyanic principles. These are destroyed by boilings, expression, etc., after which the roots become edible.
8. In the Philippines tapioca is used for preparing alcoholic fuel.

Other Uses of Starch

1. Starch is used in the manufacture of anhydrous dextrose for intravenous injections.
2. Starch is used for manufacturing glucose to be used in pharmaceuticals and confectionery.
3. Other industrial uses of starch are in the manufacture of textiles, paper, dry cell batteries, etc.

YAMS

(Dioscorea spp.)

Origin and History

Yams are starch-rich tuber producing plants of the hot tropics. Most of the species are of food value whereas a few are drug-yielding ones. The word 'Yam' is usually used for the cultivated species.

Yams are believed to have originated in **South East Asia** from where some species got transferred to the New World. There is so much difference in the Old World and New World species of Yams that they are believed to have had a divergent evolution. The species were domesticated in **Asia, Africa** and **tropical America** separately.

Radiocarbon dating has shown the oldest domesticated species to be 10,000 B.C. old or even more. These were found to be associated with mesolithic Hoabinhian ceramics found in southeast Asia. The most important and widely spread domesticated species is ***D. alata*** (white Yam). It spread through Asiatic tropics and Africa and was carried across the Pacific. It was carried to America via Atlantic Ocean as well.

In the New World, Yams are believed to have been domesticated first on the borders of **Brazil** and **Guyana** and then in the **Caribbean islands**.

There are 600 species of which about 10 species are food crops. *Dioscorea of Yams* are cultivated as food crop in West Africa, parts of southeast Asia and Oceania and in the Caribbean and adjoining areas of tropical America.

Botanical Description

Yams belong to the monocot family **Dioscoreaceae**. They are climbers. They bear alternate or opposite leaves which are very broad. Their roots are highly tuberous. They mostly bear unisexual flowers or remain flowerless. The flowers are very small and occur in axillary spikes. The female flower bears a tricarpellary gynoceium enclosed within two whorls of three perianth each. The flower is epigynous. The fruit is a three chambered capsule or berry. Each chamber has two winged seeds. The male flowers occur in an elongated inflorescence. They are present in several axillary long panicles. The perianth is arranged in two whorls of three each with six stamens and a rudimentary ovary enclosed inside.

Cultivation

Yams have numerous varieties. The differences between them is all due to the differences in the types of tubers produced by them. The plants are usually propagated vegetatively by means of tubers or stem cuttings. In most of the species in which only one large tuber is produced by the plant the top or 'crown' of the tuber, which contains buds or eyes, is cut off and is used for planting. In some species in which several small-sized tubers are produced the entire tuber is used for planting. They are normally grown on mounds or ridges. The stem is very weak and, therefore, it is provided with supports. The stem twines around the support in a clockwise or anticlockwise manner. It takes 8—12 months before the tubers are harvested.

The tubers are very large. Some varieties produce tubers up to six feet long. The tubers are usually buried deep in the soil. They grow vertically downwards. The tubers are sometimes branched. They are produced singly or in a cluster.

Yams are rich in starch and proteins. They also contain a poisonous alkaloid, **dioscorine**. The amount of the alkaloid is small in the edible species and is, therefore, destroyed on boiling.

Yams can be stored for several months.

Important Species

1. ***Dioscorea alata.*** White Yam or Abiatic Yam. It is most widespread and grows in abundance in Vietnam, Cambodia, Laos and other tropical regions of Asia.
2. ***D. rotundata***- White Guinea Yam. It is grown in high-rainfall zone of West Africa and the West Indies.
3. ***D. cayenensis***-Yellow Guinea Yam. It is also known as the Twelve Month Yam.
4. ***D. eseulenta***-Lesser Yam or Chinese Yam. It is grown in Pacific islands, tropical Asia viz.,Malayasia, Burma, Sri Lanka and parts of West Africa.
5. ***D. bulbifera***— Potato Yam. It is grown in tropical Africa, the West Indies and parts of Asia.
6. ***D. trifida***- Cush-cush Yam. It is grown in West Indies and tropical America.

Uses of Yams

1. In the hot tropics where rainfall is widespread yams along with tapioca constitute the staple food for millions of poor people.
2. Yam tubers are peeled off and cooked in various ways. Boiling and mashing is very common. They are also eaten after roasting or frying.
3. Since yams store well the tubers of even wild species are very useful during famine conditions.
4. Wild yams from arid regions of South Africa and Mexico are collected for extracting steroidal saponins for the manufacture of drugs.

Stem Vegetables

POTATO

(Vern. Alu)

(Solanum tuberosum)

Origin and History

Potato is generally known as white or '**Irish**' **potato**. It is one of the chief vegetables and one of the most important food plants of the world.

The wild potato tubers have toxic alkaloids and are bitter in taste. Potatoes grew in a wild state on the high slopes of the **Andes mountains in Peru and Chile in South America** where there are weed, wild and cultivated potatoes, the whole group forming a gigantic complex with hybridization and polyploidy. The plant has been under cultivation for the last 2500 years. The small, not very good tubers of potato soon gave way to large tasty tubers. The ancient Indians of Peru cultivated potato on a large scals. The tubers were dried, grounded into flour and were baked or soaked in water until they softened.

Potato is believed to have been domesticated in the high plateau of **Bolivia-Peru** in the general region of **Lake Titicacea** (Fig. 6.7). The area abounds in wild and cultivated diploids, cultivated tetraploids and triploids. The tetraploids are named *Solanum andigena, S. tuberosum* subsp. *andigena.*

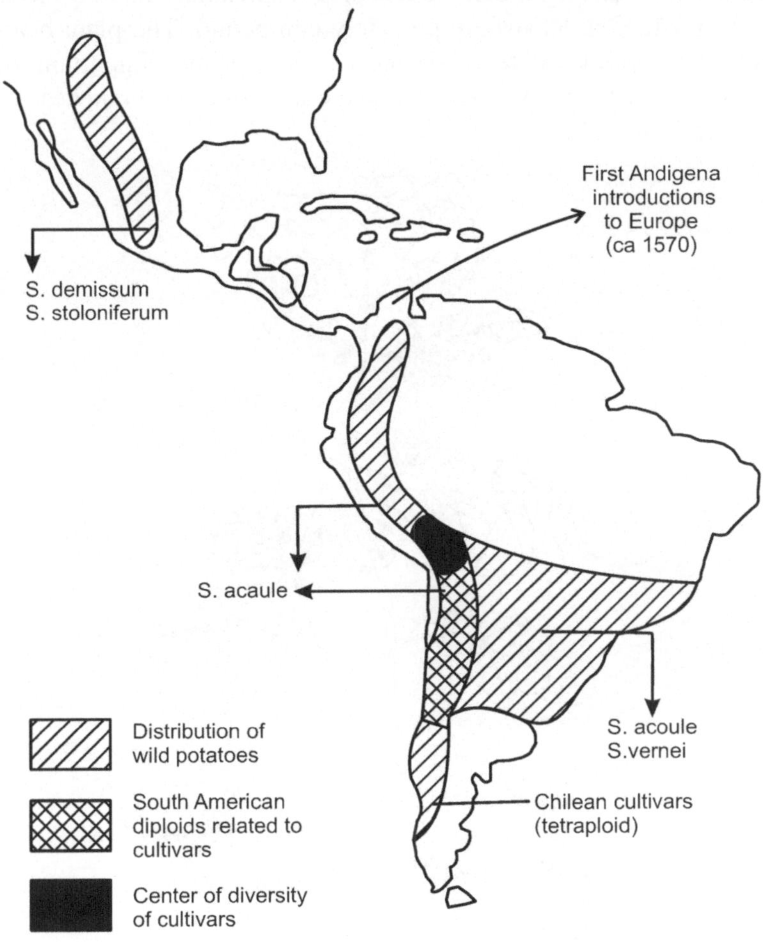

Fig. 6.7. Centers of origin of potato in America.

Spaniards carried potatoes to Europe in about 1580 A.D. Soon it spread to Portugal, France, Germany, Switzerland and to the rest of Europe including the Great Britain. It became a very important crop of Ireland. The Irish immigrants introduced potato to New England (North America) in 1719 A.D. The English men were, however, very much prejudiced against potatoes for about 200 years. They called the tubers evil and unfit for consumption by Christians since there was no reference to potatoes in the Bible. Europe today accounts for 90 per cent of potatoes, North America 5 per cent, and the rest of the world 5 per cent of the total production of the crop.

Botanical Description

The potato plant belongs to the family **Solanaceae**. The plant is an annual. It is weak-stemmed, very much branched, and spreading plant. It has a height of only 2 to 3 ft. The leaves are pinnately compound. The plant bears small white, yellow, or purple flowers and small berries. The most important structures in a potato plant are the swollen tips of the underground branches. They are called tubers (Fig. 6.8).

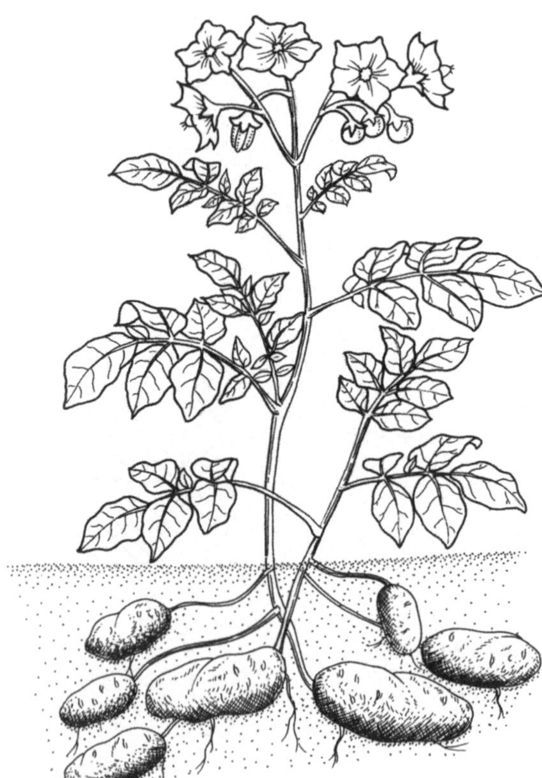

Fig. 6.8. A plant of potato.

The tubers bear a number of eyes which are groups of buds borne in the axils of aborted leaves. Each eye has a central bud which is surrounded by smaller lateral buds. Potatoes are propagated vegetatively by means of tuber pieces having at least one eye on them. The tuber should be cut in pieces at right angles to the main axis to remove the inhibiting effect of the apical bud.

The tuber skin (periderm) varies in colour, texture and thickness. Inside the skin there is a narrow cortex of small starch-rich cells, a ring of vascular budles and a large pith composed of starch bearing cells. The starch grains swell up on cooking and break up the thin cell walls to give the characteristic 'mealiness' of a cooked potato. The potatoes have 19 per cent carbohydrates, 2 per cent proteins, 0.1 per cent fat and 78 per cent water.

Ecological Factors

Climate–Potatoes are usually cultivated in a cool and moist climate.. In India potatoes thrive best on the cool hills of the north. In plains it is grown as an irrigated crop. As a matter of fact potatoes can grow in many climates all the world over except in low tropical regions. Due to their hardiness and capacity to mature quickly they can be grown in extremely cold regions of the north up to 64° NL and at altitudes upto 8000 ft. The plants favour a mean annual temperature of 40 to 50°F.

Soil-Potatoes grow in many different kinds of soils. Rich light soil is best for potatoes. They can also grow very well on poor soils.

Cultivation and Harvesting

The potatoes are sown in carefully prepared seed beds which are spaced 1-2 ft. apart. The cut seed pieces (eyes) are planted about 10 cm deep and at intervals of 20 cm by hand or by drills.

Harvesting is done by hand or by special tractor-drawn or self-propelled apron diggers that dig the tubers and strain them free of soil and debris. The harvested tubers are kept in a warm shady place to encourage the development of the protective outer corky layer (periderm). Potatoes are also sent directly to the markets from the field. The tubers are usually stored in cold storage godowns. Poor farmers store them by spreading the tubers on the floors of cool rooms. The tubers are covered with sand wherever available to keep the temperature low.

Uses of Potato

1. Potatoes constitute one of the important foods of certain countries. In India they are used throughout the year as the main vegetable.
2. A number of delicacies like potato cutlets, chips, 'papar' etc. are prepared from them.

3. A considerable portion of the crop is fed to the stock in European countries.
4. Potatoes have a number of industrial uses. Potato starch is particularly suitable for sizing paper and is more frequently used for finishing fine cotton goods. Pre-gelatinized starch is used for the production of moulded articles and poster paints. Oxidised starch is included in printing inks for cotton and linen goods. Dextrins and adhesives are produced from acidified starch. Dextrose and commercial glucose are prepared by acidic hydrolysis of starch. Alcohol may be obtained by the fermentation, by yeast, of a cooked mass of potatoes, followed by distillation. Fresh potato tubers are an important source of vitamins B and C. The former is significant for curing of beri-beri and the latter for curing scurvy.

AROIDS

(Vern. Arvi, Ghuiya, Taro, Dasheen)

(Colocasia antiquorum; C. esculenta)

Origin and History

There are about one thousand horticultural varieties of these plants. They all belong to the genus *Colocasia* and are commonly called **aroids**. They are believed to have originated in **south eastern Asia** and in the **Pacific islands**. **Spier** (1951) believes that the cultivated varieties originated in Assam or Upper Burma. Taro spread early to Japan, the West Asia, and Northern Africa. In Africa it was a common food in slave-trading time. Later they were carried to the New World along with the slaves. They are grown in almost all the warm tropical countries.

Botanical Description

The plants belong to the family **Araceae**. The stem is modified into a large fleshy underground rhizome, root stock or a corm (Fig. 6.9). It bears nodes, internodes, scaly leaves, axillary and terminal buds. The leaves are very large and peltate with long petioles. The leaves are 4 to 6 ft. in length. The plants do not flower and are, therefore, propageted by means of tops and buds. They contain about 25 per cent starch, 68 per cent water, 3 per cent sugar, 2 per cent protein and little fat.

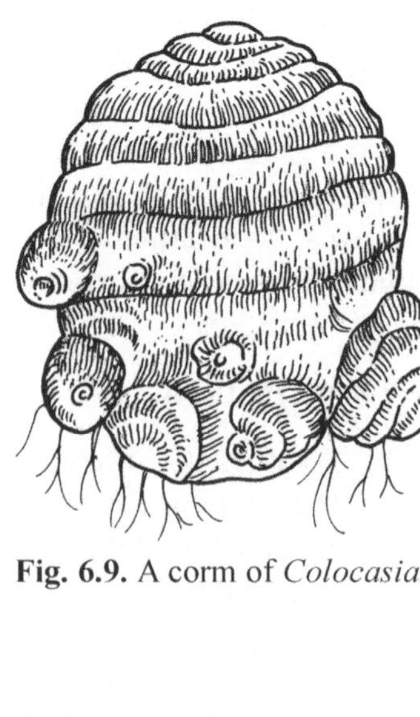

Fig. 6.9. A corm of *Colocasia.*

Cultivation

They grow in moist tropical climates. They grow well in wet soil. They take a very long time to mature. "Arvis" are cultivated in irrigated fields. They are planted by hand. They are harvested by digging up the whole plant.

Uses of Aroids

1. The rhizomes or corms of aroids form an important food for millions of tropical people particularly in the Orient. The fleshy corms are eaten boiled, baked, mashed or dried and pulverized to destroy the biting **calcium oxalate** crystals which they cotnain in abundance. The starch of aroids makes the fleshy meal which is very much liked. It is easily digestible. It contains carbohydrates and proteins in quantities which are more than those of potatoes.
2. They are used in preparing chips, canned **poi**, flours, infant foods and breakfast foods. The poi is a famous highly palatable dish of the Hawaiians. It is a thin pasty fermented mass of taro starch. It can be made in cakes or can be baked or toasted.

ONION
(Vern. Pyaj)
(Allium cepa)

Origin and History

Onion has not been found in the wild state. It is believed to have originated in the **Western Asia**, where they were cultivated in ancient times. The Egyptian workers building the Great pyramid, about 4,800 years ago are reported to have eaten onions and garlic. Pictures of onions are carved on ancient monuments of Egypt. The Bible records that Israelies complained because Moses did not bring onions for food when he led his people out of Egypt (1500 B.C). The British priests, the Druids used onions in their ceremonies about 2,000 years ago. In India onions and garlics were used in medicine in 600 B.C. References to onions are also found in the writings of Greek and Roman authors during two or four centuries B.C. Later Spanish colonizers spread the onions to the 'Indies' soon after the year 1500, and then to Mexico and South America. Other Europeans introduced onions to the United States and Canada. Plant breeders have since then developed many new varieties. Onions are now cultivated throughout the world.

Botanical Description

Allium cepa belongs to the **Liliaceae** family. The stem is reduced to the form of a **disc**. It bears a large number of concentric leaves above and fibrous roots below. The leaves are specially modified into food storage organs. These constitute the main part of the bulb. The bulb may be large or small. They may be white, yellow, pinkish red or light brows. The bulbs have strong or mild flavours. The terminal bud grows out into a leafless scape which bears an attractive umbellate inflorescence at the top. Many cultivated onions have clusters of tiny bulbs at the tops of their stalks instead of flowers and seeds. The bulbs are used for growing new onion plants.

Cultivation and Harvesting

Onions are grown in temperate regions having cool and moist climates. They are also grown in tropical regions. They prefer sandy soil.

The seeds of onions are drill-sown. The small seedlings are usually transplanted in carefully prepared seed-beds. Sometimes small bulbs formed at the tip of onion stalk are used as **sets** for growing new plants.

Harvesting is usually done by hand or by ploughing. Onions are spread out in the open fields to dry them before they are put in gunny bags for storage purposes.

Uses of Onions

1. The fleshy leaves of onion are rich in sugar. They are usually eaten raw and as salad. They are a very common ingredient of cooked vegetables.
2. Their characteristic pungent flavour is due to the presence of an acrid volatile oil **allylsulphide**. They are, therefore, very commonly used as flavouring agents.
3. Onions are also used as medicinal plants. The onion juice is believed to be antibacterial and hence it is taken in the case of digestive disorders and infections.
4. Onions are pickled by keeping them in vinegar etc. The most expensive 'midget' pickled onions are made from bulbs that grow upon the scape. They are small and have strong flavour.

II. HERBAGE VEGETABLES

CABBAGE AND ITS ALLIES

(Vern. Patta Gobhi, Ganth Gobhi, etc.)

(Brassica oleracea)

Origin and History

There are many varieties of cabbage and cabbage-like plants. They all are believed to have arisen from sea cabbage, which still grows wild in **northern**

Europe, **Asia Minor** and in regions near the **Mediterranean Sea**. The sea cabbage also called **colewort** began to be cultivated 4,000 to 5,000 years ago. **Kale** and **Collards** (var. ***acephala***) are closest to these wild froms.

The cabbage (var. ***capitaita***) is of two kinds—**loose-headed** and **hard-headed**. The loose-headed cabbage (Savoy cabbage) does not have stem. It consists of a short stalk on the top of which a large number of crinkly leaves form a bunch or head. This type is supposed to have developed directly from wild cabbage near the Mediterranean Sea. Greeks and Romans ate loose-leaf cabbage raw or boiled, just as we do today. The leaves of hard-headed cabbage are wrapped tightly around each other and around the tip of the stalk or '**heart**' (Fig. 6.10). It is assumed to have evolved from wild cabbage among the mountains of Asia Minor more than 2,200 years ago. The seeds were carried to northern Europe by warrior bands of people called Celts. German and their neighbours also grew them in ancient times. It was grown in England in 1300's and was introduced into Scotland by British soldiers before 1650. **Jacques Carlier** grew hard-headed cabbage in Canada in 1540. Today it is worldwide in distribution. North Europe and North America produce huge quantities of head-cabbage.

Botanical Description

The plants belongs to the family **Brassicaceae**. It is a biennial. Due to extreme condensing of the nodes the large glabrous leaves ball together a few nodes above the base to give the typical compact, swollen 'cabbage'. The loose-headed cabbage has, however, loosely arranged leaves. The hard-headed cabbage has several horticultural varieties : (*i*) Round green heads, (*ii*) Egg shaped or cone shaped, (*iii*) Flat headed varieties, (*iv*) Round red cabbage etc. In the red cabbage the red colouring is confined-to the epidermal layer. Cabbage has 91 per cent water. It contains some starch, protein, vitamins and minerals.

Fig. 6.10. A hard-headed cabbage.

Cultivation

It grows well in cool and moist conditions. It likes fertile heavy soils though it can grow on all kinds of soils.

The seedlings are transplanted in prepared seed beds by hand; They are spaced apart. The plants are sown shallow but when the head develops the soil is raised all around the plant.

The heads are harvested by hand. They are either stored in special storage godowns or are sent to the market directly.

Uses of Cabbage

1. It is eaten raw or cooked. Steaming is better than boiling since the nutrients are retained.
2. It is a common stock feed.
3. Fermented pieces of cabbage with salt contain lactic acid. This is known an sauerkrant and is a favourite food in Asia, Germany. America and Russia.

CAULIFLOWER (var. *botrytis).* It is commonly known as 'phool-gobhi' in our country. Cauliflower is the best of the cabbages. It was grown in West Asia in the pre-Christian era. It was introduced in northern Europe towards the end of the Middle Ages. Cauliflower and Broccoli plants have short thick erect stems which are surrounded by leaves. The stem bears a large inflorescence at

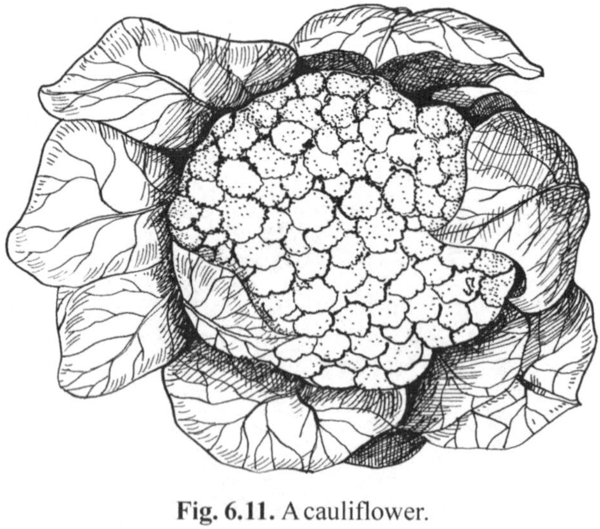

Fig. 6.11. A cauliflower.

the top (Fig. 6.11). The inflorescence consists of large number of abortive flowers borne on thick hypertrophied branches.

It is a very common vegetable of the temperate region. The inflorescence is eaten raw or is cooked as a vegetable. It is easily digestible. It is widely cultivated in the northern region of our country during winter season.

Other cabbages are **Kohlrabi** (var. ***gongylodes***: vern. Ganth Gobhi), **Brussels Sprouts** (var. ***gemmifera*** : vern. Buttam Gobhi) etc.

III. FRUIT VEGETABLES

CUCURBITS

(Cucumbers, Gourds, Pumpkins Squashes and Melons)

CUCUMBERS

(Cucumis spp.)

Cucumbers belong to the family *Cucurbitaceae.* They were first grown in **northern India**, about 4,000 years ago. One of the cucumbers, ***C. sativus*** (vern. **Khira**) has been referred to in the ancient writings of the Hebrews, Egyptians, Greeks and Romans. The traders spread the cucumbers to the West and the East in the next 2,000 years. Chinese are reported to have cultivated cucumbers about 2,100 years ago. A Roman emperor took a strong fancy for cucumbers and ate them daily. Though the ancient cucumbers were warm climate plants, the cool climate varieties were grown in France before the year 800 A.D., and they became common in England during the 1300's. Cucumber seeds were carried to America in the fifteenth century.

The plant is a trailing vine. The stem is rough. It bears yellow axillary flowers. The fruits are usually elongated. The fruits range from few centimetres to about 50 centimetres in length.

Cucumbers grow in hot climates and in sandy soils. They are harvested by hand.

Cucumbers are of three kinds:

(i) ***C. sativus*** (vern. Khira)

(ii) ***C. meloutilissimus*** (vern. Kakari)

(iii) ***C. anguria*** (gherkin of West Indies).

Cucumbers are eaten raw, pickled, or cooked as vegetables. They are very popular as salad. Gherkins of West Indies are used mainly for preparing pickles.

GOURDS

(Vern. Lauki, Kadoo, etc.)

Gourds belong to the family **Cucurbitaceae**. They were among the first food plants grown in America. Natives near the **northern coast of Peru** grew them almost 5,000 years ago and called them bottle gourd ***(Lagenaria vulgaris)***. Its vines have heart shaped leaves, white flowers, and fruits which are sometimes very long (Fig. 6.12). The bottle shaped gourds were used as bottles, cups and bowls. Bottles made of gourd as old as 3,500 years have been discovered. The seeds of this plant were carried to Polynesia. It then spread to eastern Asia, India, Asia Minor and then to Europe. Young fruits were stuffed with rice and chopped meat and were boiled. Arabs and Turks still eat stuffed green gourds.

The gourd is a very common vegetable of India.

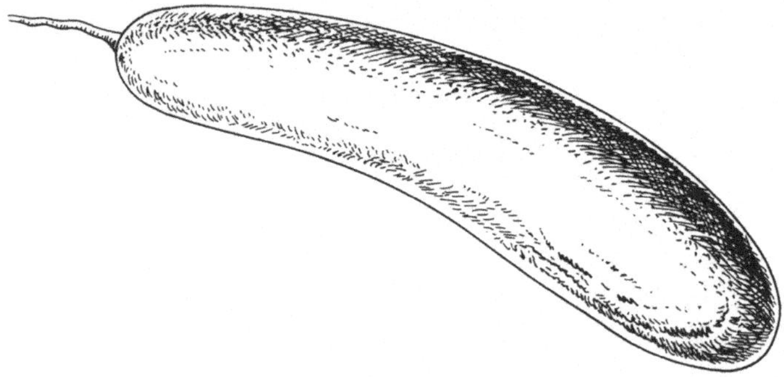

Fig. 6.12. A fruit of bottle gourd.

PUMPKINS AND SQUASHES
(Cucurbita spp.)

Pumpkins and summer squashes are also gourd fruits. They belong to the genus ***Cucurbita*** and the family **Cucurbitaceae**. They are special varieties of the yellow flowered gourd which is native to **North America**. They are believed to have developed from yellow gourds that were taken to Central America during ancient times. The traders carried the seeds of the new plants to U.S.A. and eastern Canada in the north and Peru in the south. Since none of the five cultivated species have been found in the wild state it is obvious that they have been in cultivation since the ancient times.

The plants are trailing vines with a very large root system. The stem is angular and the large ovate and lobed leaves are prickly. The flowers are large and yellow. Insect pollination is quite common. The fruits are of various shapes and sizes (Fig. 6.13). The cultivated species are as follows:

(i) ***Cucurbita pepo*** (pumpkin)
(ii) ***C. moschata*** (squash)
(iii) ***C. mixta*** („)
(iv) ***C. maxima*** („)
(v) ***C. ficifolia*** („)

Fig. 6.13. A fruit of pumpkin.

C. pepo includes field pumpkins, summer squashes, vegetable marrows, acorn squashes, etc., Young pumpkins and squashes are eaten after they are boiled, mashed, or fried. The mature fruits are baked, canned or are used as a feed of the livestock. Pumpkin seeds are also edible. They are fried and sold in the market.

MELONS

(Muskmelon and Watermelon)

Melons are usually not called vegetables. There are two kinds of melons : muskmelon ***(Cucumis melo***; vern. **Kharbuja**) and watermelon ***(Colocynthis citrullus***, originally ***Citrullus vulgaris;*** Vern. **Tarbooj**). They belong to the family **Cucurbitaceae**.

Watermelons are believed to be native of **Africa**. They are in cultivation for over 4,000 years. They were carried from Egypt to Asia and Europe about 2,000 years ago. Muskmelons (cantaloupes) occur in wide range of forms. They originated in Asia (Persia) from where the traders carried them to Europe in 1500's. Muskmelons have been reported to be in cultivation in Egypt in 2400 B.C.

The plants are trailing vines. They are annuals. The stem is pubescent and angular. The leaves are large and lobed. The flowers are yellow and monoecious. The fruits are usually large (Fig. 6.14). Both yellow-fleshed and red-fleshed bitter and sweet types are known. The melons are hot weather plants. They grow well in warm sandy soils.

Fig. 6.14. A Water melon.

Watermelons are very sweet in taste. They are eaten raw or are iced as dessert. Due to their very high water content they are consumed by men and animals during the hot summer to quench thirst. They are eaten as salad. The seeds of muskmelon are called '**magaz**'. They are used in preparing sweetmeats. They are cooked in fat and salted to be eaten like salted peanuts. The rind of watermelon is sometimes made into 'pickles'. In Russia, the watermelon juice is fermented to prepare a beverage. A variety of watermelon called '**citron**' or '**preserving melon**' contains plenty of pectin because of which it is used in preparing jams and jellies.

TOMATO

(Vern. Tamatar)

(Lycopersicum esculentum)

Origin and History

Though tomato derives its name from **tomati** or **tomatl** which are the names used by the Aztec Indians of Mexico, the tomato is native of **Peru**. Wild tomatoes were very smaller than some of the cherries. In course of time the farmers started cultivating the plants which had largest fruits. The seeds from such plants were sown year after year which ultimately resulted in new varieties with larger and tastier tomatoes.

Fig. 6.15. Fruits of tomato.

The seeds were introduced into **Central Mexico**, where it was first domesticated. Spanish explorers carried the seeds of tomato to Spain before they carried the potatoes. Later, it spread to Italy, France, and other countries. An Italian wrote in 1544 that this fruit, brought to Italy in our times is eaten in the same manner as mushrooms—it is fried in oil with salt and pepper. Italians also boiled tomatoes with salt and spices, and used them as sauce in other foods. Italians developed smooth varieties with strong skins and 'meaty' flesh from flattish tomatoes which had thick, rounded ribs. English people used tomatoes as medicine to cure stomach-aches, sore eyes, cuts and to prevent fits, etc. The French also grew tomatoes for ornamental purposes. They considered them 'objects of affection' and, therefore, called them **'love apples'**. **Thomas Jefferson** was one of the first Americans to grow tomatoes for food, in about 1781. Americans, however, considered these 'love apples' to be poisonous. It is interesting to note that a man in 1820 became famous by eating tomatoes on the court house steps in Salem, New Jersey. Tomato now is of great importance next only to potato and sweet potato.

Botanical Description and Cultivation

The tomatoes belong to the family **Solanaceae**. The tomato plant is erect or trailing in habit. The stem is coarse and is very much branched. The fruit is a berry. The fruit may vary in size from very small to very large. They are usually rounded but some may be pear-shaped (Fig. 6.15). The fruits contain largely water but also have sugars, acids and vitamins. The plant can be grown as a perennial or as an annual. The plant is grown in hot as well as cool climates. The fruits are picked by hand.

Uses of Tomatoes

1. Tomatoes are eaten raw or as salad. They are rich in vitamins particularly vitamin C.
2. They are cooked as vegetable.
3. The pulp is processed into juice, canned tomatoes, ketchup, sauce and tomato pastes.
4. Green tomatoes are used for 'pickles'.
5. The seeds of tomato give an antibiotic, tomatine.
6. Tomatoes are also consumed by animals.

EGGPLANT

(Brinjal; Vern. Baingan)

(Solanum melongena)

Origin and History

Eggplants are native to **Northern India**, where wild eggplants still grow as weeds. Seeds were carried to China about 1,500 years ago, where it became a popular food. Eggplants were taken to Europe by Arab traders during the Dark Ages. They were grown in Spain before the year 1,200 A.D., and they soon spread to other parts of Europe. Eggplants were introduced into Brazil before the close of seventeenth century. They are now extensively grown in the warmer regions of both hemispheres. Today it is the fourth ranking vegetable in Japan.

Botanical Description and Cultivation

The eggplant belongs to the family **Solanaceae**. It is an erect branched bushy annual plant with hairy or spiny stems that grow to a height of 4 feet. The leaves are large and thick. The fruit is shaped like a ball, a melon, or a huge egg (Fig. 6.16). Wild varieties are green, but cultivated ones may be green, yellow, brown, gray, dark purple or even white.

They grow in warm climates and require a high temperature; In temperate climates they are sown in hotbeds and then transplanted.

Fig. 6.16. Fruit of an eggplant.

Uses of Eggplant

1. Eggplant fruits are very common vegetable in India and many other countries.
2. The fruits are used as pickles.
3. Preparations like 'Pakoras' etc. are made from the slices of the fruits.

QUESTIONS

1. Which is the chief vegetable yielding family? Name three vegetables belonging to the family along with their botanical description.
2. Describe in detail distribution, the origin, the botanical features, and climatic conditions for the cultivation of potato.
3. Describe the botanical features of sweet potato and tapioca.
4. What are aroids? Give the botanical description of any one of them along with its uses.

7

Food Plants (Contd.)

FRUITS

In the preceding chapters we have seen that the commercial importance of majority of the food plants lies in their fruits or in the seeds which the fruits contain. The cereals, legumes, nuts all bear edible dry fruits. Many edible fleshy fruits have been classed as vegetables. The term 'fruit' is commonly used for only those which are eaten without cooking.

Morphologically the fruits are of three kinds—**simple**, **aggregate** and **compound**. The fruits can be broadly classified into **temperate** and **tropical** types. Whereas the temperate fruits are cultivated, the tropical ones are semi-cultivated and almost grow wild. The temperate fruits do not constitute a staple food whereas the tropical fruits supplement the diet of the people of the region. The important temperate fruits are **pome fruits**, *viz.,* apple, pear, medlar, quince: **stone fruits**, *viz.,* apricot, cherry, peach, plumes and prune; **gourds**, *viz.,* melons, watermelons, **berries**, *viz.,* blackberries, raspberries, blueberries, cranberries, currant and gooseberries, mulberry, strawberries and grapes, etc. The important tropical fruits include citrus fruits, *viz.,* oranges, pomelo, lemon, lime, kumquats, citron, banana, custard apples, date, fig, guava, litchi, loquat, mango, olive, papaya, pineapple, pomegranate, etc. Only four fruits, two from each group are described below.

APPLE

(Pyrus malus)

Origin and History

The apple is believed to have originated in **eastern Europe** and in the **Caucasus Mountain** of **Western Asia**. It is still found growing in wild forms in the latter region. They have been cultivated for over 3,000 years. Seeds of apple have been found in the remains of Lake Dwellers. Today apple is extensively cultivated in the temperate zones and has over 6,500 horticultural forms. Each colonists introduced it into America. The chief apple growing regions are North America, Western Europe, Australia, New Zealand and South Africa.

Botanical Description and Cultivation

Apple belongs to the family **Rosaceae**. It is a perennial tree with a height of 40 feet or so. A tree of apple can produce fruits for as long as 100 years. The orchards are, however, renewed with better varieties and younger trees about every forty or fifty years.

Apples are usually propagated by grafting selected varieties on the stock grown from seeds, cuttings or layerings. All the branches bear leaves but only the short shoots, (spurs) bear white and pink coloured flowers. Pollination is by insect particularly bees. The fruit is surrounded by fleshy receptacle (thalamus) which is the edible part. Such a fruit is called pome (Fig. 7.1).

Fig. 7.1. Apples.

The fruits are green, yellow, golden or red coloured. The ripe fruit has more sugar and less starch and acid. Each variety develops the characteristic flavour and aroma. Apple fruits have 84% water, 0.3% protein, 0.4% fat, 11% sugar and 0.5% acid.

Apples can grow in many different kinds of soils and climates. Lime-rich soils are best for giving maximum yield. Perfectly ripe fruits are plucked from the tree.

Uses of Apple

1. Apples are very delicious fruits. They are often eaten raw.
2. Apples are also cooked, dried or canned.
3. The apple juice is fermented by yeast to produce alcoholic product called hard cider.
4. Alcoholic product is fermented by acetic acid bacteria to produce vinegar.

5. Fresh apple juice is a common drink.
6. Jams and jellies are prepared from apple.
7. The wood of the apple tree is very hard. It is used as firewood and for tool handles.

GRAPES

(Vitis vinifera)

Origin and History

Grapes grow wild in the temperate regions of **Asia**, **Africa**, **Europe** and **America**. The ancient wine grape is one of the oldest cultivated plant. It is native of **South-Western Asia**. It has been cultivated for 6,000 years in Egypt. Bible has several references of grapes. The grapes were carried to Greece, Italy and then to the rest of Europe. Later they found way to the New World. Now a number of horticultural varieties of grapes have been developed almost all from the ancient wine grape. They are now widely grown in the temperate regions. The chief grape-growing regions are Europe, U.S.A., Argentina, Chile, Australia and South Africa.

Botanical Description and Cultivation

Grape belongs to the family **Vitaceae**. The plant is a climbing vine. It is woody and climbs with the help of tendrils. The vine bears large, palmate, alternately arranged leaves. The flowers are small and fragrant. The fruit is a berry. The fruits are always produced in bunches. The fruits have very high sugar content. They have a very thin skin. The grapes are usually propagated by cuttings.

Grapes grow very well on well-drained, humus rich, loose soil. The Meditterranean type of climate is best suited to its growth.

Clean cultivation is practised in the vineyard. The plants are pruned to a length of 3 to 4 feet, so that the vines support large clusters of quality fruit. Grapes are very susceptible to fungal infection.

Uses of Grapes

1. Grapes are consumed raw since they are very sweet in taste.
2. Grapes are extensively grown for preparing good quality of wine and brandy.
3. They are dried to give raisins and currants.
4. Wild and other grapes are used in preparing jams and jellies.
5. The waste products of the grape industry are used as stock feed and fertilizers.

6. The fruits are used in preparing acetic acid, cream of tartar, a fixed oil and tannin.
7. Grape plants are beautiful to look at and are sometimes grown for ornamental purposes.

SWEET ORANGE

(Vern. Mausambi)

(Citrus sinensis)

Sweet orange is a tropical fruit. It is one of the citrus fruits which include besides sweet orange ***(C. sinensis)***. lemon ***(C. limon)***, sour orange ***(C. aurantium)***, limes ***(C. aurantifolia)***, mandarin orange ***(C. reticulata)***, citrons ***(C. medico)*** and many others.

Origin and History

Sweet orange like most of the citrus fruits, wild and cultivated, originated in **South-Eastern Asia**, probably **China**. Sweet orange was domesticated between 1,500 and 1,000 B.C. Soon after it was carried to India. In the early part of the fifteenth century the Genoese traders carried it to Europe. The Portuguese and Spanish explorers carried it to the New World in the sixteenth century. Now the chief orange-growing countries are Brazil, Spain, Italy, Palestine, Mexico, U.S.A. and India.

Botanical Description

Citrus species belong to the family **Rutaceae** the plant is of small height (up to 20 ft). It is an evergreen tree. It has large number of spines. The leaves are thick and have winged petioles. The flowers are white. The fruits are rounded. They are usually yellow, brown or orange in colour. Botanically it is a hesperidium. The placental hairs are juicy and edible. The fruits contain 5 to 10% sugar, 1 to 2% citric acid and vitamin C.

The plant grows well in warm dry climate. The soil should be fertile and well irrigated.

Uses of Sweet Orange

1. The juicy portion of the orange is consumed fresh. It is relished very much because of its sweetness.
2. Orange juice is a very popular drink.
3. Oil is extracted from the rind. The essential oil is used in medicine, in the perfume and soap industries, and for flavouring.
4. The plants are also grown for ornamental purposes.

MANGO

(Mangifera indica)

Origin and History

Mango is as important to tropical people as is apple to the temperate people. It is of greatest importance amongst the tropical fruits. Mango has a very old history. It has been under cultivation for nearly 6,000 years. All the species of *Mangifera* have originated in **southern Asia**. They extend from northern India to the Philippines and New Guinea. They have a lot of diversity in the Indochina-Malaysia-Indonesia region (Fig. 7.2). Mango is extensively grown in India, Malaysia, Indonesia, Africa and tropical America. Innumerable horticultural varieties of mango are cultivated throughout the tropics.

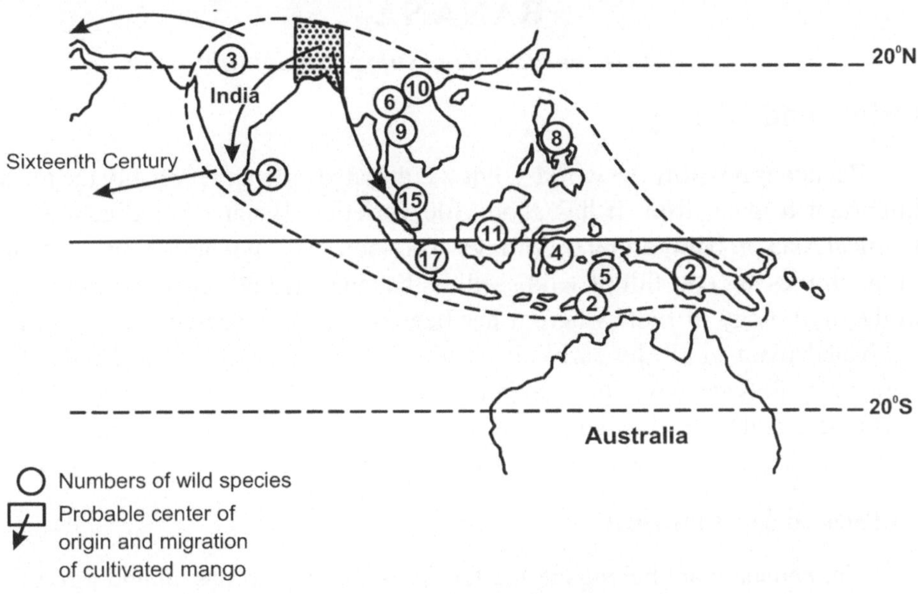

Fig. 7.2. Distribution of wild ***Mangifera spp***. and origin of ***M. Indica***.

Botanical Description

The plant belongs to the family **Anacardiaceae**. The plant is a branched evergreen tree. It grows up to a height of 90 feet. The leaves are thick. The tree bears small pink flowers. The fruits are large and fleshy. They are drupes in which the edible part is the pulpy mesocarp. There are a number of varieties of mango, the fruits of which vary greatly in size, shape, quality and flavour. The pulp is very sweet when ripe and may be orange, yellow or red in colour. Unripe fruits can be collected and ripened during storage.

Uses of Mango

1. Mangoes are a very good source of food. They are very much liked by the tropical people and are eaten fresh.
2. They are sometimes cooked as vegetable.
3. They are used in preparing pickles, 'chutney', sauces, etc.
4. Dried pieces of mango are used throughout the year as spice for cooking vegetables. The dried powder called 'amchur' is also similarly used.
5. Mango is also used in preparing jams which are canned.
6. Mango juice is added to milk to give drinks called mango shake.
7. Mango pulp is dried partly to prepare 'ampapar'.

BANANA

(Musa paradisiaca, subsp. sapientum)

Origin and History

Banana is possibly the world's oldest cultivated crop and probably the most important tropical fruit. It has about 300 varieties. Banana is indigenous to **tropical Asia** probably **India** or **Malaya**. The south-east Asiatic countries falling in the tropics are still full of-wild seedless bananas. It finds several references in the mythology of India where it has been treated as a sacred fruit. It was a cultivated plant in all the early civilizations. It was spread to all the Pacific regions by Polynasions. The Arabian traders spread it to the whole of Africa. Portuguese and Spanish colonizers also spread the plant to various places lying in the tropics. Banana reached West Indies in 1500 A.D.

Botanical Description

The banana plant belongs to the family **Musaceae**. It has an underground stem, a fleshy rhizome, and an erect 'stem' composed of sheathing bases of petioles with a crown of large leaves at the top. The plant attains a maximum height of 30 ft. and is, therefore, one of the tallest herbaceous plants. The leaves are 8 to 12 ft. long and 2 ft. in width.

The plant bears a terminal inflorescence. It is actually borne on a stalk which develops from the root stalk and grows through the hollow stem to come out at the top. The inflorescence is pendulous. It contains clusters of flowers called 'hands' or 'combs' covered by bright purpulish spathes. The latter fall down when the fruits mature. The fruit is a seedless berry. There are about 12–20 bananas ('fingers') in one cluster (Fig. 7.3).

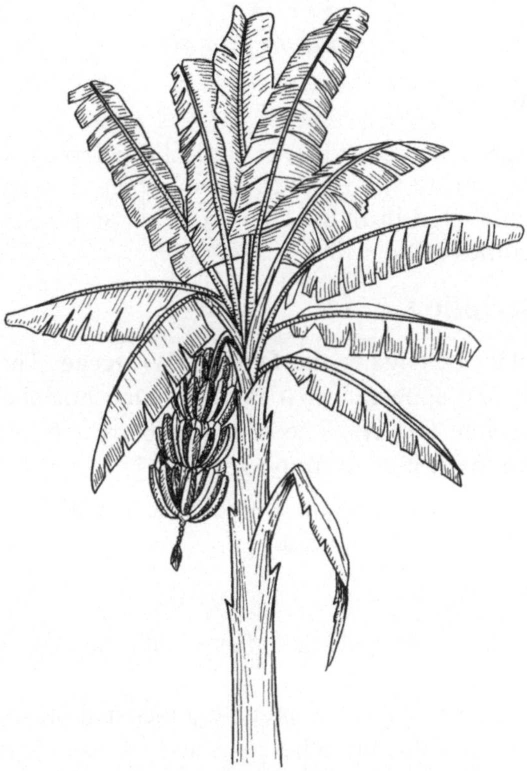

Fig. 7.3. Portion of a banana plant with an inflorescence.

Cultivation

Banana being a seedless plant is propagated by means of rhizomes or suckers which are planted in holes dug about 20 ft. apart. The weaker shoots are removed. The plant grows very rapidly and gives a very high yield.

The fruits are cut by hand when they are still green.

Uses of Banana

1. Bananas contain plenty of carbohydrates and some fats, proteins, minerals and vitamins. They are, therefore, highly nutritious. Bananas are usually eaten raw. They are very delicious to eat.
2. Green bananas are cooked as vegetable.
3. Dried green bananas give flour which is used in preparing soups etc.
4. Slices of ripe fruits are dried to give banana 'figs'
5. Bananas also serve as a stock feed.
6. Banana sheath fibre is coarse and has good strength. It is used in cottage industry for making ropes, handbags, etc.

GUAVA

(Psidium guajava)

Origin and History

Guava is indigenous to **Brazil**. This tropical fruit has been under cultivation for centuries. It has spread almost all over the tropics. Though it grows wild in the tropics it is cultivated in India, Cuba, Florida and California (U.S.A.) to produce good qualities of the fruit.

Botanical Description

The gauva plant belongs to the family **Myrtaceae**. The plant is a small tree with a shrub like appearance. It is very much branched. It bears small leaves and large white flowers. The fruit is a berry. The plant bears a very large number of yellow coloured fruits. The pulp is variously coloured.

The fruit contains sugar, pectin, vitamins A, B and C, and ascorbic acid. The plant is usually grown from seed.

Uses of Guava

1. Guava is a sweet, juicy and highly flavoured fruit. It is, therefore, delicious to eat raw.
2. Guava is used extensively to prepare jellies and preserves, etc.
3. Guava powder is added to other jams and jellies to make them firmer in consistency.

LITCHI

(***Litchi chinensis***)

Origin and History

Litchi is native of **southern China**. It has been under cultivation in south-eastern Asia for over 2,000 years. Litchi is cultivated in India, China, South Africa and some other parts of the tropics. It is now also grown in Florida and California.

Botanical Description

It belongs to the family **Sapindaceae**. The plant is a small tree of a height of 35 to 40 ft. It is highly branched and bears a large number of shining leaves. The plant is propagated by air layering. The fruits are produced during summer season. The characteristic fruits of litchi are borne in loose, terminal clusters. They are round or oblong and are 1" or 2" in diameter. The fruit is a one-seeded nut. The pericarp is spiny and is bright red in colour. The single-seed of the fruit is surrounded by a whitish fleshy "aril" which is edible.

Uses of Litchi

1. The flesh of the fruit is very sweet and juicy and is, therefore, eaten raw. It is a great delicacy in India and China.
2. Dried Litchis are also edible and are usually used for export purpose.

PAPAYA

(*Carica papaya*)

Origin and History

Papaya has originated in the **West Indies** or **Mexico**. It is found wild and is naturalized through most of the tropical America. It was introduced in India in 1598. It has since then spread through the Pacific Islands, where papaya is now very common. The chief papaya-growing regions are West Indies, India, Malaysia, Hawaiian Islands, California and Florida.

Botanical Description

The plant belongs to the family **Caricaceae**. The tree is a large herb which may attain a height of 25 ft. The stem is unbranched. The stem is soft and bears a crown of large leaves (Fig. 7.4). The plant is dioecious. It bears yellow flowers. The tree has a life of 3 to 4 years only and starts bearing fruits when it is 10 months old. The fruits are berries. They are produced in hundreds on the stem just below the leaves. They are large, hollow and pyriform shaped.

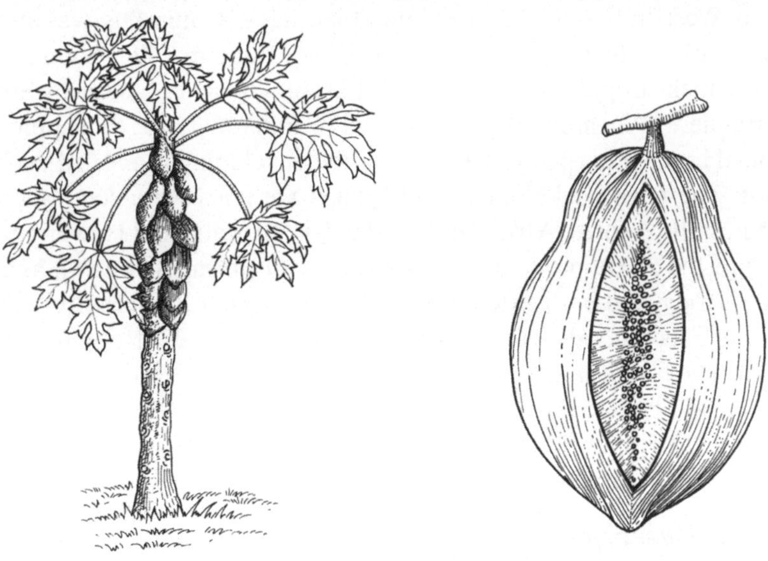

A B

Fig. 7.4. Papaya. A. Plant; B. Fruit.

Uses of Papaya

1. Ripe fruits with yellowish-orange pulp are usually eaten fresh.
2. They are used in preparing juice, pickles, jellies, etc.
3. Green fruits are cooked as a vegetable.
4. The plant including the fruit yields latex which is used in the manufacture of chewing gum.
5. The latex contains a digestive pepsin like enzyme **papain** which can hydrolyse protein.Papain is, therefore, used for tenderizing meat. Pieces of green papaya are added to meat to soften it during cooking.
6. Papain has also medicinal value.
7. The pulp of papaya is used in the manufacture of cosmetics.

PINEAPPLE
(Vern. Ananas)
(Ananas comosus)

Origin and History

The pineapple is native of **tropical South America** where it is still found in wild forms. Small-fruited, seed-beating pineapples are found growing in Brazil and Paraguay. The Guarani Indians who domesticated the wild forms spread the pineapple throughout lowland South America east of the Andes. It also spread to West Indies in the pre-Columbian days. Columbus was presented with pineapple in the Carribean island of Guadeloupe in 1493. Soon pineapple was spread to the tropical Asia, Africa, the East Indies and Polynesia by Spaniards and Portuguese. Pineapple was introduced in India in 1548. Pineapple was so much liked by the Europeans that they started developing improved varieties of the plant. The improved English and French varieties were soon adopted by West Indies, the Azores, Africa, India, Malaya, Australia and Hawaii (accounts for 3/4th of the world production). The variety of pineapple known as Smooth Cayenne is today the most extensively cultivated variety, both for fresh fruit and for canning. Its smooth leaves, with no prickles, makes it popular with the cultivator. The fruit is large, juicy, fine flavoured and cans well. This variety was introduced into France from French Guiana or Cayenne in the earlier part of the last century. The plant flourishes in tropical climate where rainfall is 50 inches and the temperature is 60° F to 90°F.

Botanical Description

Pineapple belongs to the family **Bromeliaceae**. In nature the majority of 1300 species grow as epiphytes on trees of South America. The roots of the plant are generally associated with fungal mycelium in a mycorrhizal association. ***Ananas*** is a terrestrial plant. Several varieties of pineapple are cultivated throughout the world. They differ in the taste of the fruit and in the colour of the flesh. The plant is a short lived biennial or perennial herb. It bears a large number of spirally arranged leaves which are quite compact. The leaves are fleshy and long with a sharply pointed tip. The inflorescence is terminal in position and bears several sessile reddish purple flowers. The berry-like fruits developed from individual flowers fuse together to form a composite sorosis fruit. The fruit is rich in sugar.

The ripe fruits are highly nutritive and delicious. They contain besides sugar, vitamins A and C, citric and malic acid.

The plants are propagated by vegetative means. The vegetative structures are suckers or small shoots called **"slips"** or bulbils. The plant has a very short stem and a rosette of long stiff leaves. The leaves are about 3 ft. in length. Their margins are prickly and the tips are spiny. The stem bears a large inflorescence—an oblong head of flowers.

There is a tuft of leaves at the tip of the inflorescence. The latter matures into a large multiple sorosis fruit called pineapple. It is formed by the fusion of all the parts of the inflorescence, *viz.,* the individual fruits, bracts, sepals, petals and the axis of the inflorescence itself. The large fruit is seedless in cultivated varieties. The fruit contains sugar (15%), a digestive enzyme—bromelin, fruit acids, vitamins and minerals.

Cultivation

Pineapple can grow in poor dry sandy soils in hot climates. The plants are very easily propagated. In wild state they are propagated usually by means of seeds but in commercial cultivation the suckers are planted usually in rows. When the first fruit is cut, the plant branches to give a second crop of two fruits. After they are cut, the branches again divide to bear four smaller fruits.

Fig. 7.5. Pineapple.

The fruits are harvested only when they are fully ripe so that the characteristic flavour is developed. Unripe fruits are usually cut when they are to be transported to long distances. Harvesting is done by hand. In Hawaii, long conveyor belts fitted to trucks are used for harvesting. The fruits are carried by the conveyor belt to the boxes which are kept on the truck.

Uses of Pineapple

1. It is eaten raw because of its juicy flesh and delicious flavour.
2. In the realm of canned fruits the pineapple has made great strides in recent years and there is probably more canned pineapple produced in the world today than that of any other fruit.
3. Pineapple juice is also canned extensively. It is also used in beverages.
4. There are a variety of uses of the waste products of the canning industry. The fruit 'shells' are used as a cattle-feed.
5. Fiber obtained from leaves of pineapple is processed on jute machinery. The fibre can be used for the manufacture of fabrics for curtains, furnishing, etc. In Asian countries like China, Taiwan, the Philippines the leaves are a source of a very strong fiber. The fibers are flexible and are resistant to water.

Some other Fruits

Sr. No.	Common name	Botanical name	Family
1.	Anar	*Punica granatum*	Puniaceae
2.	Amla	*Phyllanthus emblica*	Euphorbiaceae
3.	Bel (wood apple)	*Aegle marmelos*	Rutaceae
4.	Cheeku	*Achras zapota*	Sapotaceae
5.	Custard apple	*Anona squamosa*	Anonaceae
6.	Fig	*Ficus carica*	Moraceae
7.	Jack fruit	*Artocarpus integrifolia*	Moraceae
8.	Jamun	*Syzygium jambolana*	Myrtaceae
9.	Phalsa	*Grewia asiatica*	Tiliaceae
10.	Shah toot	*Morus alba*	Moraceae
11.	Strawberry	*Fragaria sp.*	Rosaceae

QUESTIONS

1. Name 5 fruit yielding plants along with their botanical names. Give the morphology and method of cultivation of any one of them.
2. Discuss the origin, distribution and botany of *Pyrus malus*.
3. Give the botanical name of the following :
   (i) Custard apple
   (ii) Pineapple
   (iii) Papaya
   (iv) Grapes and
   (v) Cheeku
4. Describe the botanical features of a fruit yielding plant belonging to the family Bromeliaceae.

8

Spices and other Flavouring Materials

The edible materials which are used more for their aromatic flavour than for their food value are known as spices. The flavour and aroma are due to essential oils, which are organic volatile compounds of relatively small molecules. The spices serve mainly three purposes: they do not let the diet to be monotonous, they disguise the stale flavour of food-materials and increase the pleasure of eating and they help to cool the body by increasing the rate of perspiration. Spices were sometimes considered so important that they were used in beverages, in medicine and even in lieu of money. Spices are generally derived from different parts of the plants like shoots, stems, leaves, flowers, seeds, etc.

It is very difficult to classify the spices. They are generally divided on the basis of the part or parts of the plant that give spices. These may be as follows:

A. Spices obtained from underground parts.
B. Spices obtained from the bark.
C. Spices obtained from the flower buds and flowers.
D. Spices obtained from the fruits.

Spices Obtained From The Underground Parts

GINGER
(Vern. Adarakh)

(Zingiber officinale)

Origin and History

It has originated in **tropical south-east Asia**. China and India were the first countries to grow it. It was later carried to Asia Minor. It was introduced in Europe early in the Middle Ages. It is now widely grown in the tropics. The chief ginger-growing countries are India, China, Japan, Sierra Leone, Indonesia, Malaysia, Queensland (Australia), Jamaica and other islands of the West Indies.

Botanical Description

Ginger belongs to the family **Zingiberaceae**. It is a perennial herb. It bears a large scaly rhizome. The rhizomes are pale yellow in colour. They contain starch, gums, an oleoresin and an essential oil. The stem attains a maximum height of 3 ft. It is surrounded by the sheathing bases of large leaves which are borne alternately on the stem. The stem bears at its tip yellowish flowers in a spike inflorescence. The fruit is a thin walled three valved capsule. It contains several small seeds.

The presence of **gingerin**, a non-volatile **oleoresin** is responsible for the pungent taste, whereas an essential oil gives the aromatic odour.

Cultivation and Storage

It is a plant of wet tropics and grows in regions of high rainfall and high temperature. It prefers a partly shaded place. The plant grows well in rich moist soil.

Whereas in China ginger is prepared by drying, cleaning, boiling, peeling, scraping and then boiling again in a sugar solution before they are packed, it is prepared by cleaning, peeling and drying in the sun in other countries. The former type is called **preserved** or **green ginger** and the latter type is called **dried** or **cured ginger**.

Uses of Ginger

1. It is used as a condiment.
2. It dilates the blood vessels, and thus causes a feeling of warmth. It increases the rate of perspiration and thus helps in lowering the body temperature.
3. It is widely used in preparing soups, puddings, pickles, cookies and curries to give flavour and pungency.
4. It is used in preparing beverages.
5. It is of medicinal importance. It is used in digestive disorders.
6. The distilled oil is used in perfumery.

TURMERIC

(Vern. Haldi)

(Curcuma longa)

Origin and History

Turmeric is obtained from the underground rhizomes of the plant. It belongs to the ginger family **Zingiberaceae**. The plant is believed to have originated in **southern Asia** (Cochin, China, Indonesia, etc.) The plant has been under cultivation since the remote past. The plant is not known in the wild state.

The plant is cultivated in the tropics. The chief turmeric-growing regions are India, China, Sri Lanka, Indonesia and West Indies. In India it is cultivated in almost all the states particularly in Tamil Nadu, Andhra Pradesh, Maharashtra, Bengal and Orissa. It also grows wild in the shaded valleys and slopes of the Western and Eastern Ghats.

Botanical Description

The plant is a perennial herb. The stem is a underground rhizome. The rhizome is short, thick, blunt and coarse. It bears fibrous adventitious roots. The rhizome bears groups of leaves. The leaves have long petioles and are 2 feet or more in height. The plant bears an inflorescence which is 4 to 7 inches in length. The flowers are pale-yellow and are covered by a group of pinkish bracts. The fruit a capsule, is rarely produced.

Turmeric is highly aromatic. It has a characteristic musky taste and odour.

Ecological Factors

The crop requires a warm and moist climate. In heavy rainfall tracts on the west coast, it is grown unirrigated; in other areas, it is cultivated under irrigation. It can be grown from almost sea level to the altitude of 4,000 feet. The wild types are seen at higher elevation. Well drained, fertile, sandy and clayey, medium black, red or alluvial loams, rich in humus and uniform in texture are highly suited for its cultivation. It is commonly grown on soils that can stand heavy irrigation.

Cultivation

The methods of preparatory tillage, manuring, sowing, irrigation for turmeric are about the same as for ginger. For sowing both round and finger-like rhizomes are cut into $1\frac{1}{2}$ to 2 inches long pieces, each having at least one sound bud. The rhizomes are planted three inches deep and 9 to 15 inches apart in either direction. The crop is sown from April to August depending upon the region of cultivation.

The leaves of the plants turn yellow and begin to dry in December to February. Harvesting is done from January to April. A small portion of the crop is left unharvested for about a month for the supply of seed material for the new crop.

Curing–The stalks of turmeric are cut a few days before the rhizomes are dug out. The harvested rhizomes are separated into smaller pieces, if necessary. They are then steeped in water in a large earthen pot or iron pan kept over a furnace: Dried turmeric leaves or pieces of gunny are spread over the pan and the water is boiled for about an hour. The softened rhizomes are then taken out, spread on the threshing floor and are exposed to sun for 10-15 days for complete drying. The bits of adhering roots and scales are removed by hand and the dried produce is either disposed of, or is polished by rocking it in suspended bamboo basket containing small stones.

Uses of Turmeric

1. It is extensively used as a dye which is not fast. The rhizome produces a yellow-orange dye, which is still used for dyeing purposes in India and China. It readily dyes cotton, silk and wool fibres, without using a mordant.
2. It is an important flavouring agent.
3. It is used for colouring butter, cheese, pickles, mustard and other foodstuffs.
4. The spice is a common constituent of curry powders used in cooking vegetables, meat, fish, etc.
5. It is also used medicinally in pains (external applications) and as a stimulant (internal administration).
6. It is used as a cosmetic in some places.
7. It is very sacred to the Hindus who use it as an offering on religious and ceremonial occasions.

Other Spices

Zedoary is similar to turmeric. It is derived from the rhizomes of ***Curcuma zedoaria***, which is grown in India on a large scale. It is used for flavouring curries, drinks, etc. It is also used in medicine, cosmetics and perfumery. Its tubers are used as a source of starch.

Wild Turmeric or Yellow Zedoary. The plant is ***Curcuma aromatica***. It grows wild in India. The rhizomes is yellow coloured and the inner flesh is orange-red. It is used as a dye and in medicine.

Mango Ginger grows wild in India. The rhizomes of ***Curcuma amada*** have the flavour of mangoes. They are used in pickles and in medicine.

Angelica is obtained from ***Angelica archangelica*** of the **Umbelliferae**, It is indigenous to Syria. It is now chiefly grown in Germany. The whole plant is highly aromatic. It is used for flavouring cakes, candy and beverages. It is used in perfumery and in medicines. Young stems and petioles are candied in syrup for decorating the candy with bright green colour as well for flavouring it.

Galangal is obtained from two species—***Alpinia officinarum*** (lesser galangal) and ***A. galanga*** (greater galangal). The rhizomes are aromatic with a pungent taste. It is used for flavouring drinks and in medicine.

Horse radish is obtained from the large roots of ***Armoracia lapathifolia*** (***A. rusticana***) of **Brassicaceae**. It is a native of south-eastern Europe. It is used as a condiment.

Sarsaparilla is obtained from the dried roots of wild species of *Smilax* of the family **Liliaceae**. The plants are native to American tropics and subtropics. They are trailing vines. It is used as a flavouring material.

Spices Obtained From Barks

CINNAMON

(Vern. Dalchini)

(Cinnamomum zeylanicum)

Origin and History

The plant belongs to the family **Lauraceae**. The spice is obtained from the bark of the tree. The plant is indigenous to **Sri Lanka**. The plant has spread to southern India, Burma, Malaysia, South America and the West Indies. The plant is now extensively grown in the tropics. Sri Lanka is the chief producer of commercial cinnamon.

Another kind of cinnamon, called Saigon cinnamon (**C. *loureirii*)** grows in Vietnam, Combodia, etc. Its coarse bark is very much in use in China, Japan and U.S.A.

Botanical Description

The plant of ***C. zeylanicum*** is an evergreen small tree. It may attain a height of 50 feet. The tree is very much branched. The cultivated plants are, however, very small. They are encouraged to become bushy and to form suckers (shoots from the roots) by pruning the young shoots almost at the ground level. The long and slender suckers yield the commercial cinnamon. The plant bears tough and broad aromatic leaves which are of a length of six or seven inches. The plant bears terminal or auxiliary inflorescences. Small yellow flowers are borne on short branches of the panicle. The fruits are fleshy drupes.

Cultivation

The plants are grown in plantations under semi-wild conditions. The trees usually flower in May and the fruits mature in two to three months time. The shoots (suckers) of a length of six to seven feet are usually cut during the rainy season when the peeling of the bark is easy. They are cut twice in a year. The new shoots are allowed to grow for two years when the shrub is again cut back to the roots. The bark is carefully removed by cutting into two long strips of three-feet lengths. The bark is collected and covered with canvas to ferment them partly. The outer portion of the bark (cork) is then removed off and the remaining strips are slowly dried. During the drying process the strips get rolled into, 'quills'. The dried quills are fitted one inside the other before they are sent to the market. The quills may be pulverized to produce the ground spice.

Uses of Cinnamon

1. The spice is extensively used for flavouring vegetables, meat, sweets and other foodstuffs. It is a constituent of curry powder.
2. It is used in perfumes, incense, dentifrices, gum and candy.
3. Cinnamon oil distilled from green leaves is rich in eugenol. It is used in the manufacture of perfumes, etc.
4. The oil derived from the waste of the bark-peeling process is used in medicine as a carminative, antiseptic and astringent.

Other Spices

Cassia (Chinese cinnamon) is derived from the bark of ***Cinnamomum cassia***, a native of Burma. Cassia bark and cassia oil are used for flavouring purposes, in medicine, soap, bakery, confectionary and perfumery. Cassia of inferior quality is also derived from the barks of ***C. tamala, C. burmannii, C. oliueri*** and ***C. massoia***.

Sassafras. It is obtained from the bark and chipped wood of ***Sassafras albidum***. It also belongs to the family **Lauraceae**. It is native of eastern North America. It is a flavouring material. It is used for flavouring medicines, beer and other beverages, soaps, tobacco, perfumes, gum, etc.

Spices Obtained From Flowers or Flower Buds

CLOVES

(Vern. Laung)

(Eugenia or ***Syzygium aromatica)***

Origin and History

Cloves, named after the French word for nail, *le clou* are dried unopened flower buds of ***Eugenia aromatica***. The plant belongs to the family **Myrtaceae**.

They are reported to be in use in China in 300 B.C. The Romans also knew of cloves. It spread to the rest of Europe during the Middle Ages. The plant is believed to be a native of **Moluccas**. In the 17th and 18th centuries the trade in cloves was a Dutch monopoly. Later French grew them in Mauritius and Reunion in 1770. From there they were carried to the British East African colony of Zanzibar and the neighbouring island of Pemba. The plant is now widely cultivated in the tropics. The chief clove-growing regions are Tanzania (Zanzibar) (which accounts for 90 per cent of the total world production), Indonesia, Mauritius, and the West Indies.

Botanical Description

***Eugenia aromatica* (*E. caryophyllata*)** is a small evergreen tree. The plant is usually of 25 to 40 feet in height but the wild ones may attain a height of 70 feet. The tree is very much branched into forks which carry large number of lateral branches in an upward direction. The whole tree gives a conical and very symmetrical appearance. The leaves are borne in opposite manner. They are highly aromatic due to the presence of oil glands.

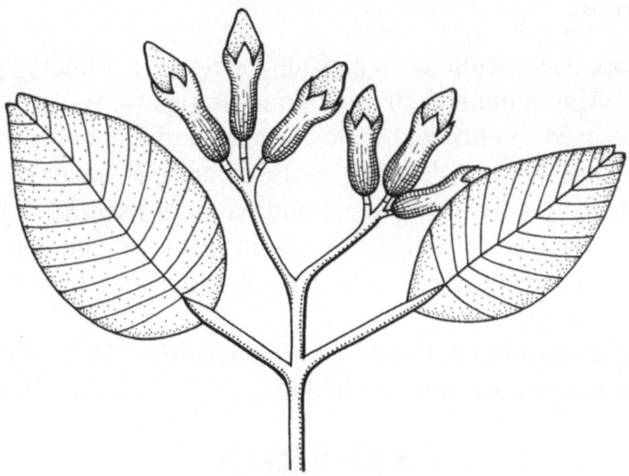

Fig. 8.1. Part of a twig of clove plant.

Wild plants have clusters of crimson flowers which are produced in terminal cymes. The flowers are in groups of three (Fig. 8.1). The flowers have cylindrical bases. The sepals of a flower form a fleshy tube which encloses an inferior ovary. The sepals divide into a four-toothed structure. There are four crimson petals. The petals fall down when the flower opens. There are numerous stamens and only one carpel (Fig. 8.3). The ovary matures into a cylindrical one seeded drupe. Cloves are highly aromatic and have a flavour.

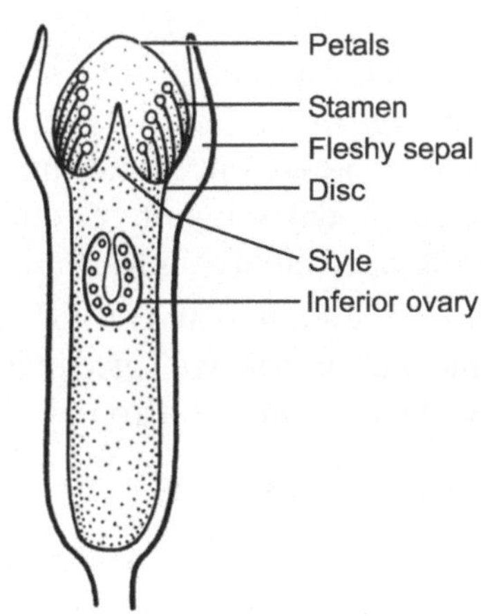

Fig. 8.2. Section of bud of a clove.

Cultivation

The clove tree grows well in the lowland forest of the wet tropics. It requires a rainfall of 50" to 70". It thrives in loamy soils. It can grow up to an altitude of 2,000 feet.

The tree is very difficult to grow. The germination of the seeds is very slow and takes 5 to 6 weeks to complete. Nursery seedlings used for large scale cultivation also grow very slowly. The tree starts flowering at the age of 8 or 9 years but good yield of cloves is obtained only when the trees are 20 years old. The trees stop giving yield at the age of 60 years or more. The buds are removed while they are still dull green in colour. The buds are picked by hand. They are then 'steamed' and dried before they are sent to the market.

Uses of Cloves

1. Cloves are used whole or in a ground state as a culinary spice. It is used for flavouring a number of vegetarian preparations.
2. They are used in curries, pickles, sauces and ketchup.
3. Clove oil prepared by distillation of flower buds, stems of fruits contains **eugenol**. It is used in medicine and as an antiseptic. It is used for the relief of toothache. The oil is also used in perfumery, in soap, in histological work as a clearing agent. It is an ingredient of many toothpastes and mouth washes.
4. They have warming and stimulating properties. They are taken in betel. Clove cigarettes are smoked in Java.

SAFFRON

(Vern. Kesar)

(Crocus sativus)

Saffron is obtained from the carpels of the flower of *Crocus sativus* which belongs to the family **Iridaceae**. The plant has been cultivated by the Greeks and Hebrews since times immemorial. It was grown in England in the Middle Ages. Today it is cultivated in the Asian tropics on a limited scale. In India, it is cultivated chiefly in Kashmir Valley.

The stigmas and tips of the styles are dried to give the saffron. It is probably the most expensive spice in the world. Seventy thousand flowers are needed to yield a pound of saffron.

Uses of Saffron

1. It is chiefly used as a dyestuff. A pinch of saffron in hot water gives a rich golden colour to the foodstuff. Rice (Pullao) and a number of favourite dishes are coloured with saffron. It is also used as a colouring material in confectionery.
2. It is used as a flavouring material.
3. It is used in perfumery.
4. It has medicinal properties. It is effective in relieving cold, cough, etc.
5. It is used by the Hindus on a number of religious and ceremonial occasions.

Spices Obtained From Fruits

CHILLIES

(Red and Green Peppers; Vern. Mirch)

(Capsicum spp.)

Origin and History

No one knows exactly as to where the ***Capsicums*** were first cultivated. They are, however, believed to be indigenous to the **tropical areas of Central America** and the **West Indies**. They are very old plants. Farmers of northern Peru grew them at least 3,500 years ago, and perhaps even more than thousand years before that. After the discovery of America and the West Indies they immediately spread throughout the tropical regions of the world. Numerous varieties, differing in habit and mode of growth and in size, shape and colour of the fruits have evolved in the course of time in different environments. Today peppers are grown in each and every tropical country of the world and have become the most important spice in a number of countries. It was introduced in India from the tropical South America in the 17th century. It is grown in all parts of our country. Andhra Pradesh, Maharashtra, Karnataka and Tamil Nadu account for about 75 per cent of the annual production.

Botanical Description

Chillies belong to the family **Solanaceae**. There are a number of varieties of chillies—sweet or bell peppers, paprikas (pimeintos), tabasco, cayenne pepper, chilli and others. Broadly speaking, all of them can be considered to belong to two species : ***Capsicum annum*** and ***Capsicum frutescens*** of which the former is more important. Some workers, however, do not consider them as two different species.

The fruits of ***C. annum*** are milder than those of ***C. frutescens*** which have fiery pungency in them. The former includes large-sized fruits often used as vegetables and salad ('sweet' pepper, paprikas, pimiento, etc.) and the latter includes the chillies of commerce (cayenne pepper, etc.) which are used as a spice. The two species are described below :

C. annum. The plants are annuals with a short growing season. They are small and bushy. They grow to a height of 1 to 3 feet. The dark green leaves are oval in shape. The flowers and fruits are borne singly in the axils of leaves. The sweet or bell peppers (Shimla mirch, etc.), and the European paprikas belong to this species. The fruits are many seeded large fleshy berries.

They vary in length from half an inch to ten or more inches. The unripe fruits are green in colour and become yellow or red when ripe. They are conical, spherical, long, or large stumpy in shape. They are very mild because either they are totally lacking in pungency (paprikas) or the pungency is limited to seeds (sweet pepper) only.

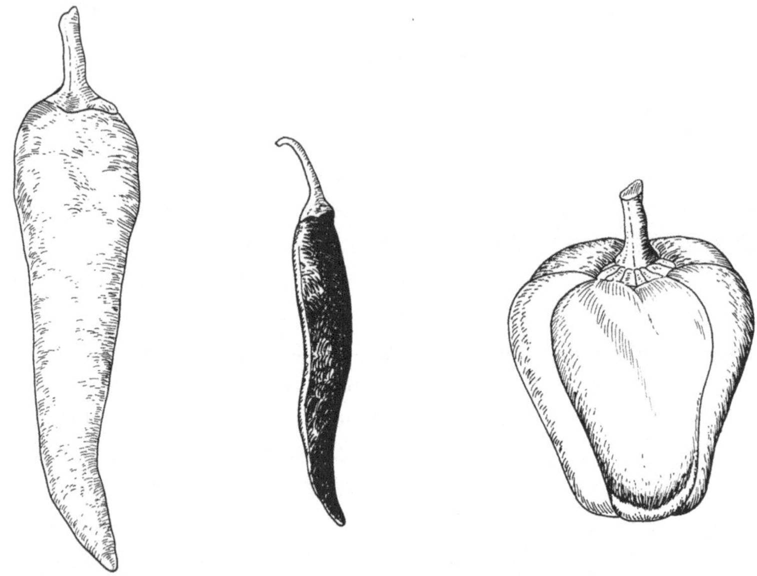

Fig. 8.3. Fruits of three varieties of ***C. frutescence***.

C. frutescens. The plants are perennial but are treated as an annual in cultivation and require a longer growing season. They are woody and taller. They can grow up to 3 or 4 years. The plant attains a height of 6 feet or more. They are tropical and subtropical plants. They include the hot chillies, cayenne pepper and tabasco (Fig. 8.3). The fruits are small elongated and conical. They are very pungent. They are usually deep red in colour when ripe.

The fruits are pungent due to the presence of a volatile compound called **capsaicin**. The fruits also contain vitamins C, A and E.

Ecological Factors

The crop is grown from sea level to an altitude of 5,000 ft. in tropical and subtropical conditions as well as cooler temperate zones with an annual rainfall of 25 to 50 inches. Very heavy rainfall during crop growth is harmful. In low rainfall tracts or when grown in hot weather, it is cultivated as an irrigated crop.

The rainfed crop does well in deep fertile, well drained black cotton soils and also in somewhat heavy clayey loams. In poorly-drained soils, plants shed leaves and turn sickly even with temporary waterlogging. Under irrigation and good manuring, excellent crops can be raised on sandy and light alluvial loams and red loamy soils.

Cultivation

The land is ploughed and harrowed three or four times. The land for irrigated chillies is laid out into beds 6 to 8 feet square, or is made into ridges 18 to 36 inches apart. The winter crop is planted in July to September and the summer crop in February to March. The seeds are first sown in nursery, and the seedlings are transplanted when five to six weeks old. Seeds taken from healthy fruits are mixed with earth and are then sown evenly in well-manured nursery beds. Seedlings are transplanted on a cloudy evening preceding anticipated rain or when it is actually drizzling.

Harvesting

Fruits of winter crop commence ripening from November onward. Four to ten pickings of ripe fruits are taken during three or four months. If there is demand for green chillies, the first picking is taken for this purpose. Ripe fruits are picked along with stalks and heaped for three or four days for the partially ripe fruits to develop the proper red colour. They are then dried in the sun for four to ten days depending upon the weather conditions and are graded for size and colour before marketing.

Uses of Chillies

1. The large green non-pungent form of *C. annum* are used as salad, and are cooked as vegetables. The thick and hollow flesh of the fruits are usually stuffed with meat, potatoes, or other vegetables.
2. The paprikas are preserved and are used in cheese preparations and stuffed olives. Powdered paprika is a common condiment in the European dishes.
3. Both green and ripe fruits of the pungent type are used as a spice. The fruits are powdered to give cayenne or red pepper. The chillies (whole dry red fruits) of commerce or the cayenne pepper is used for flavouring and colouring vegetarian and non-vegetarian preparations.
4. The chillies are used in the preparations of sauces ('chutney', curry powder and pickles). Pickles of large sized red chillies are greatly liked in our country.
5. The fruits are a rich source of vitamin C. They also contain vitamins A and E.
6. They are used in medicine as a stimulant and carminative and to prevent fever. Externally it acts as a counter irritant.
7. It is of wide use in beverage like ginger ale because of its pungency.

PEPPER
(Vern. Kali Mirch)
(Piper nigrum)

Origin and History

True pepper is the most widely used spice and has a very old history. It is believed to have originated in **south India** and **south-eastern** Asia (Indo-Malayan) region. The Greeks and the Romans used them. The Europeans developed a great liking for them during the Middle Ages. Pepper became such an important commodity of the trade in the Western Europe in the twelfth cantury that a hectic search began for a trade route to India.

The plant is chiefly grown in the Oriental tropics. It is grown in Africa, India, Malaysia, Indonesia, Thailand, the Philippine Island and South Sea Islands. In India it is chiefly grown in Kerala and Karnataka. It is also cultivated in Assam and Bengal.

Fig. 8.4. True pepper.

Botanical Description

Piper nigrum belongs to the family **Piperaceae**. The plant is a weak trailing shrub which climbs with the help of a pole or a tree. In wild conditions the plant is as tall as 30 feet. The height of the plant is, however, kept low under cultivation to facilitate harvesting. The stem has swollen nodes from which arise the adventitious roots, leaves and inflorescences. The stem is thin but becomes woody with age. Large leaves are borne in an alternate manner. The inflorescences are borne opposite the leaves (Fig. 8.4). Fifty to hundred small flowers are borne in elongated catkin inflorescences. The fruits are one-seeded spherical drupes. Matured fruits are bright red in colour. The dried fruit (peppercorn) is wrinkled and black coloured. The seed is whitish in colour. The fruit is aromatic (volatile oil) and pungent (oleoresin).

Ecological Factors

The plant grows well in a highly humid and warm tropical climate, having an annual rainfall of nearly 100 inches. It grows from sea-level to an altitude of 4,000 feet. The plant prefers shade to direct sunlight.

The plant grows best on a fertile, well drained loam soil, rich in humus. It can also be cultivated on the laterite soils of the hill slopes.

Cultivation

Pepper is generally cultivated by means of stem cuttings taken from the top of an actively growing shoot of 10 to 20 years old plant. It is planted near a supporting pole etc. The tip of the plant is removed to promote the development of lateral buds, when the plant attains a height of about 2 feet. Though the plant starts flowering and fruiting in 2 or 3 years, maximum yield is attained only after 7 years. The yield declines from 25th to 30th year. Sometimes it is also propagated by seeds.

The fruit crop is harvested from December to March depending upon the altitude of the place.

Black pepper. Black pepper of commerce is more pungent than the white pepper. For preparing black pepper the inflorescence is harvested when the fruits are still unripe and red in colour. The spikes are soaked in hot boiling water and then dried in the sun or in smoke. The dried reddish brown or black wrinkled fruits or peppercorns are rubbed off before they are sent to the market. The pepper contains various kinds of resins and a, yellow crystalline alkaloid **piperin** (4.5 to 8%). They are responsible for the pungency of pepper.

White pepper. White pepper is prepared from ripe fruits. The fruits are collected in heaps and are allowed to ferment or are soaked in running water. The pericarp and the pulp of the fruits are then removed by rubbing or pressing.

The greyish white rounded seeds are powdered to form the white pepper of commerce. White pepper is preferred to black pepper.

Uses of Pepper

1. Pepper is the most important of the spices and is used as such in the whole world.
2. It is used widely as a condiment and flavouring material for all types of foodstuffs and dishes. It is used for preserving and in the manufacture of sauces and ketchups, etc.
3. It has also medicinal uses. Its alkaloid is used in medicine.
4. Its alkaloid is used as a source of synthetic heliotrope.
5. Pepper brings about a cooling effect on the human body.
6. It stimulates the flow of saliva and gastric juices and thus helps in the digestion of food.

The Umbelliferous Fruits (Spices)

The fruits of some umbelliferous plants are economically important since they are aromatic and are, therefore, used as spices and flavourings, etc. The fruit is typically a schizocarp which splits up into two one seeded mericarps. There are ridges and furrows of the mericarp wall. The resins, gums and essential oils present in the furrows of the mericarp wall are responsible for the characteristic flavour and smell of the fruits. The important spices yielding plants are cultivated in the tropics, particularly in Northern Africa and India. They include caraway *(**Carum carvi**)*, fennel (***Foeniculum vulgare***), coriander *(**Coriandrum sativum**)*, anise *(**Pimpinella anisum**)*, cumin *(**Cumimum cyminum**)*, celery *(**Apium graveolens** var. **dulce**)* and dill *(**Peucedanum** or **Anethum graveolens**)*. Some of them are described below:

FENNEL
(Vern, Saunf)
(Foeniculum vulgare)

It is native of the **Mediterranean region**. The spice was used by the ancient Indians, Chinese and Egyptians. The plant was also cultivated by the Romans. The plant is now worldwide in distribution but is chiefly grown in India.

The plant is a tall perennial herb. The plant bears finely dissected leaves. Numerous small yellow flowers are borne in compound umbels. The fruits are oval and green or yellow-brown in colour. The wall has five prominent ridges (Fig. 8.5A).

Fennel is used as a spice in cooking vegetable etc. It is used in preparation of candy, and confection of various kinds. The fruit contains 12 to 18 per cent of semi-drying oil which is used as a carminative in medicine, flavouring material and in the manufacture of soap. It also contains volatile oil (1 to 6%) which is used medicinally and for scenting soaps. Sometimes the leaf-stalks are used as a vegetable.

CORIANDER

(Vern. Dhania)

(Coriandrum sativum)

It is indigenous to the **eastern Mediterranean**, the **Caucasus** and the **West Asia**. Coriander finds references in the very old Egyptian, Hebrew, Roman and Sanskrit literature. It grows in the wild state in Sudan and Egypt. The plant is widely grown in Europe (Holland, Hungary, Russia.), Morocco, South America and India.

The plant is a highly aromatic small perennial herb. It is, however, treated as an annual in cultivation. The plant attains a height of 2 to 3 feet. The lower leaves are broad and undivided while the upper ones are segmented. The inflorescences are typically compound umbels of small white or pink flowers. The fruits are globular and smooth (Fig 8.5B). They are yellow-brown in colour. The fruit does not break up easily into two mericarps. It has five small flaps near the apex, the remains of the five calyx lobes. While the smell of fresh fruits is unpleasant the dried fruits have very pleasant smell. The fruits contain about 1 per cent essential oil and 11 to 21 per cent of a dark brownish-green coloured fixed oil.

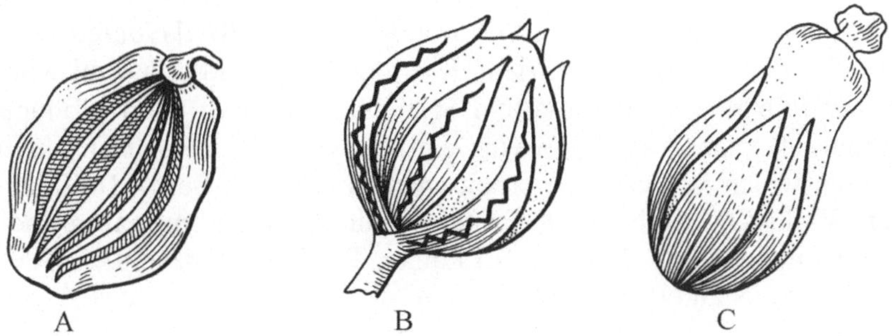

Fig. 8.5. Fruits of three economically important members of the family Umbelliferae. A. ***Foeniculum vulgare***; B. ***Coriandrum sativum***; ***C. Pimpinella anisum***.

The whole or powdered coriander is a very common constituent of curry powders. It is used as flavouring for a number of sweet and savoury dishes, 'chats' etc. The fruits are also used in the preparation of sugar candies. Coriander oil is also used in flavouring beverages, *e.g.,* whisky, gin and various liquors. It is also used in medicine and veterinary medicine.

A Fruit of arive is shown in Fig. 8.5C.

CUMIN
(Vern. Jeera)
(Cumimum cyminum)

It is also of ancient cultivation. It is believed to be native of **the Mediterranean region–upper Egypt** and **Arabia**. The plant is extensively cultivated in India, Egypt, Sudan, Morocco, Sicily, the Mediterranean islands, and Southern Europe.

The plant is a small annual of hardly a foot in height. The upper leaves are finely segmented. Small purple or white flowers are borne in compound umbels. The fruits are elongated and oval. They have narrow ridged hairy mericarps which separate easily. They contain an essential oil (4%) and a greenish coloured fixed oil (10%).

The fruits is an ingredient of curry powder. It is used in soup, cake, bread, biscuits and pickles. It is also candied. The fried fruits are used in 'dal' to flavour it. The fruits are used in medicine and veterinary medicine.

Spices Obtained From Seeds

CARDAMOM
(Vern. Illaychi)
(Elettaria cardamomum)

Origin and History

The plant is native of India. The aromatic and highly flavoured seeds had been used for centuries in the Orient. It is found in a wild or a semi-wild state in the hot rain forests of India and Sri Lanka, the two chief producers of cardamom. The plant is also grown in Central America (Guatemala in particular). Cardamom which is a very important spice is native to **South India** and **Sri Lanka**. While 80% of the world's total output of cardamom is from India the rest is obtained from Guatemata, Sri Lanka, Thailand, Tanzania, El Slavador, Costa Rica, Laos and Cambodia.

In India cardamom plant is grown in Southern states of Kerala, Karnataka and Tamilnadu. Kerala accounts for over 60% of the total production in our country.

Botanical Description

Elettaria belongs to the family ***Zingiberaceae***. It has two varieties

(i) ***E. cardamomum*** var. ***major***.

(ii) ***E. cardamomum*** var. ***cardamomum*** or var. ***minor***.

E. Cardamomum var. ***major*** grows in Sri Lanka and Kerala. It is the wild cardamom. The plant is about 10 feet in height. The stem is a pseudostem which bears large leaves. The plant bears large number of flowers, dark brown fruits, and large number of seeds.

The other variety is a cultivated one ***E. cardamomum*** var. ***cardamomum*** is a perennial herb. The underground rhizome bears numerous aerial leafy branches. The leaves are lanceolate in shape. The plant bears a very large number of flowers. The fruits are small but the seeds are highly aromatic. The perisperm of the seed is rich in starch. The volatile oil which is responsible for the aroma of seeds contain terpeneol, turpinene, cineol, sabinene and limonene.

It is a perennial-herb. The plant has a thick ginger like underground rhizome which bears adventitious roots below and aerial stems above. The aerial stems may be vegetative or reproductive in nature. The vegetative stems grow to a height of 12 feet and bear numerous alternate long (up to 3 ft.) lanceolate leaves with strongly sheathing leaf bases. The reproductive stems are leafless and usually spread out horizontally. They are very much branched. Numerous white flowers are borne in recemes in the axils of the leafy bracts (Fig. 8.6). The fruits are light brown in colour and are triangular in shape. The fruit wall is papery and easily breaks up into three valves. They contain dark-grey or black-coloured seeds.

Fig. 8.6. ***Elettaria***. A flowering twig.

There are two varieties of cardamoms, the Sri Lanka and the Malabar varieties with minor differences in their morphological features.

Ecological Factors

The plant grows only in the tropics and only at altitudes ranging from 2000 ft to 5000 ft. The mountaneous habitats have a rainfall between 250 cm and

300 cm. The rains are distributed throughout the year. The plants require protection from direct sunlight, and therefore, flourish under a canopy of other plants. The soil is rich in humus due to regular fall of leaves from the surrounding plants.

Cultivation

It is cultivated either as a pure plantation crop or as subsidiary to coffee and areca nut plantations. It is also found as a natural undergrowth in some favourable tracts of forests. The plantation is done in forests which get evenly spread out annual rainfall of 100—250 inches. It requires a deep good textured humus rich soil. It grows best at altitudes of 2,000–5000 feet under light shade conditions.

The propagation is done by means of pieces of rhizomes ('bulb') or seedlings grown in nurseries. One or two seedlings of about two years of age are planted in well manured pits of 18" × 18" × 18" in the months of June and July. Cardamoms flower when they are 3 years old.

The capsules are harvested before they are fully ripe. Harvested fruits, ripe or unripe, are dried in the sun for three or four days. Sunning partially bleaches the capsules while artificial heating retains their green colour. The seeds contain starch and 4 to 7 per cent of a volatile oil.

Uses of Cardamom

1. Cardamoms are used for flavouring curries, cakes, biscuits, pickles, sweet dishes, and for other culinary purposes.
2. They are used in medicines as carminatives.
3. The oil is also used in cooking and for flavouring.
4. Cardamom seeds are consumed whole or in a crushed form. They are used as additives in several food preparations such as curry powder, pickles, etc.

MUSTARD

(Brassica spp.)

Mustard has been described in Chapter 16 as an oil-yielding crop. Mustard seeds have a very long history and find several references in the Bible and in Greek and Roman writings. Sanskrit records of 3000 B.C. refer to mustard as in important spice.

Brassica spp. are believed to have originated in three different centres lying in the north temperate zone. ***Brassica nigra*** perhaps originated in the **Asia Minor-Iran** area from where it spread to Europe, Africa, India and the southeast Asia. ***B. juncea*** is believed to have originated in the **Central Asia-Himalayas** from where it spread to three secondary centres in India, China and Caucasus (Fig. 8.7). ***Sinapsis alba*** is believed to have originated in **Mediterranean region**. They have spread to sub-tropics and tropics as cold weather crops.

They are temperate plants but are now also cultivated in the tropical and subtropical regions. Mustard is extensively cultivated in North America, in the prairie provinces of Canada, U.K., Denmark, France, Germany, China, Japan and India. In India they are cultivated in Punjab, Haryana, U.P., Bihar, Bengal and Assam. In our country the mustard oil is normally derived from a mixture of yellow sarsoan, 'toria' and 'rai'. There is often a mixed cultivation of these plants.

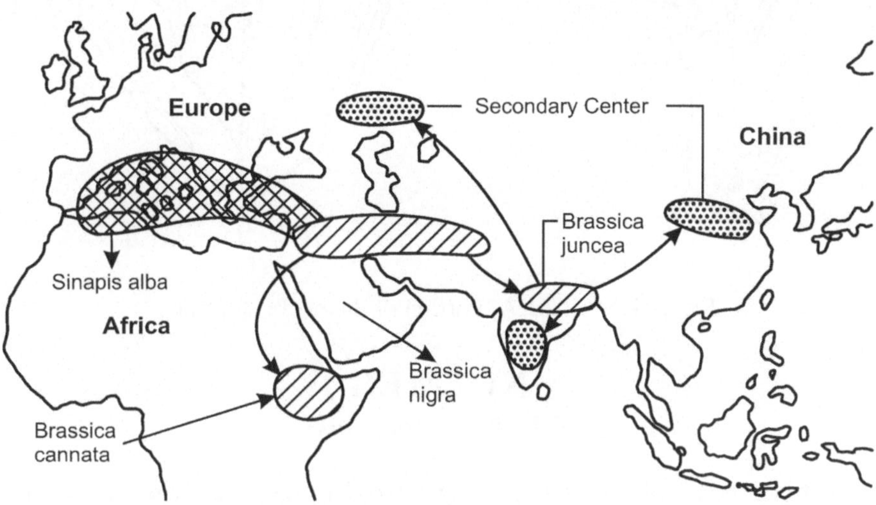

Fig. 8.7. Evolutionary geography of the mustards, ***Brassica spp.*** and ***Sinapsis alba***.

There are at least four kinds of mustards: **white mustard** ***(Brassica hirta*** or ***Sinapsis alba),*** **black mustard** ***(B. nigra*** or ***B. campestris*** *var.* ***dichotoma),*** **Indian (Brown) mustard** ***(B. juncea)*** and **yellow mustard** ***(B. campestris*** var. ***sarsoan). B. juncea*** is the most widely cultivated mustard today. The plant is an annual herb. It bears a racemose inflorescence at the tip (Fig. 8.8).

The seeds are a very important spice. They are used whole or ground in cooking vegetables and in preparing pickles, sardines, salad, dressing, etc. It stimulates the salivary glands and the alimentary canal movements. The seeds contain a fixed oil (see Chapter 16) and a glucoside, **sinalbin** (white mustard) or **sinigrin** (black mustard). The glucoside decomposes to give a nonvolatile or a volatile sulphur compound which is responsible for the characteristic sharp taste and pungent and aromatic smell. The mustard oil is a cooking oil and has a number of other uses. It is used for massaging body, in medicines and for burning lamp etc. It is also a very good lubricant.

A B

Fig. 8.8. Mustard. A. Portion of the plant; B. Fruits.

NUTMEG

(Myristica fragrans)

Nutmeg is a well known spice. It is sold in the market by the name of **nutmeg** and **mace.** Nutmeg is dried seed from which the testa has been removed. Mace on the other hand is the dried **aril** which is attached to the tip of the seed.

Nutmeg is indigenous to **Moluccas Islands** which is generally called **'Spice Island'**. Indonesia accounts for more than 60% of the world's total production of nutmeg and mace. Grenada in the West Indies, Malaysia and Sri Lanka are the other nutmeg growing countries. In India nutmeg is grown in southern states of Tamil Nadu, and Kerala and Assam.

Botanical Description

Myristica fragrans is an evergreen tree attaining a height of 30 ft to 40 ft. The plant bears dark green leaves and fleshy flowers. Inside the fruit the seeds are enclosed within the aril (Fig. 8.9). The brown coloured oily seeds are irregularly furrowed.

Ecological Factors

The nutmeg tree flourishes in hot and moist climate where temperature ranges between 25°C and 33°C and there is uniform rainfall of about 200 cm to 250 cm distributed throughout the year. The plant grows well in humus rich sandy soils.

Fig. 8.9. A twing of *Myristica fragrans.*

Harvesting

The nutmeg tree bear fruits throughout the year. Maximum yield is, however, obtained twice in a year—early winter and late winter. The plants starts flowering by the 5th year and continues to do so for even more than 50 years. A plant can bear hundreds and thousands of fruits. After ripening the fruits and the seeds are collected in baskets. The husk of the fruit is removed and the seeds are taken out.

Nutmeg seeds are dried kernels. Alternatively the aril is separated from the seed and is flattened with the help of presses before drying it up in the sun or in an oven. This type of processed aril is called **mace.** The nutmeg and mace are sold in the market by brands like West Indian and East Indian. The latter has lesser fatty oils but has a much better flavour.

Uses of Nutmeg

1. Both the nutmeg and mace are used as spices.
2. The nutmeg oil is used in the manufacture of ointments, soaps and perfumes.
3. The crushed powdery nutmeg and mace are widely used in the manufacture of biscuits, cakes, pastries etc.
4. The powdery material is also used for flavouring beverages, sweet dishes and many other food preparations.
5. Nutmeg is of medicinal value. It is carminative and has stimulative and aphrodisiac qualities.
6. Nutmeg oil is used for scenting creams and soaps.

QUESTIONS

1. Give the botanical names of the four types of mustards. Discuss the botany of mustard and also mention its uses.
2. Give the origin, distribution and botanical description of clove yielding plant.
3. What is a spice? Give a general account of any four important spices of India.
4. Give a general account of the cultivation and economic importance of spices studied by you.
5. Give botanical name, family and botany of the part used in any 5 spices studied by you.

9

Algae as Food

Algae have constituted a part of food of human beings, from the ancient times. 'Materia medica' published in China several thousand years ago have references to ***Laminaria*** and ***Gracilaria*** as food plants. The early Japanese ate ***Porphyra*** along with rice and, therefore, cultivated this alga on a large scale. Seaweeds were widely used by the Polynesians. In nineteenth century about 75 species of algae were already used as food plants in the Hawaii island. In the past few centuries carragheen ***(Chondrus crispus)*** has been widely used in Western Europe.

Algae are a potent source of carbohydrates, proteins, vitamins, minerals etc. Algae are responsible for ninety per cent of the total photosynthesis taking place on earth. The phytoplankton which comprise the floating algae and the other algae are eaten by fish and other aquatic animals.

An account of the edible algae is given below:

1. Cyanophyta

Nostoc communae has been used as an edible alga in China for a very long time. It is known as **Yuyucho** in China. Other blue green algae which are edible are species of ***Microcystis*** and ***Anabaenopsis*** and ***Spirulina***. They are eaten by man and fishes alike.

2. Chlorophyta

Several green algae are edible. Important chlorophycean algae which are edible are ***Chlorella, Ulva lactuca, Monostroma, Codium sp.*** and ***Enteromorpha compressa***.

Chlorella pyrenoidosa contains a high percentage of protein (30%), carbohydrates (30%), lipids (15%) and ash (5%). The proteins in ***Chlorella*** species have amino acids which are fit for human consumption. These species also contain vitamins B, C, K, provitamin A, thiamine, biotin, riboflavin, folic acid, niacin, inositol, and pyridoxine; several minerals like sulphur, phosphorus, manganese, potassium, iron, copper, zinc and cobalt and their salts.

Ulva lactuca is used in Scotland for preparing soup and salad. ***Monostroma*** is known as ***Aonori*** in Japan. It is used as salad.

3. Phaeophyta

Several brown algae are edible. The young stipes of ***Laminaria saccharina*** are eaten in Japan. The preparation from the alga is known as **'Kombu'.** Several kinds of Kombu are marketed in Japan. They are termed, 'Kombu Black', 'Kombu White', 'Pulpy Kombu' and 'Sweet Cake Kombu'.

The other edible algae are ***Alaria sp., Sargassum sp.*** and ***Nereocystis sp.***

Brown algae are rich in carbohydrates (57.04%), proteins (6.15%) and fats (1.56%). Thirty to seventy per cent of the seventeen aminoacids present in the proteins are digestible. The carbohydrates contain digestible laminarin (36%), mannitol (5 —37%), cellulose (1—9%) and algin (50%). Brown algae are also rich in minerals and vitamins.

4. Rhodophyta

The alga ***Porphyra tenera*** is the most popular edible alga in Japan. It is known as **amanori.** Its preparation is known as **'Asakusa-Novi'**. Large areas are being used for cultivating this species in Japan as well as U.S.A. ***Rhodymenia palmata*** is used as food in England, Scotland and Ireland. ***Porphyra spp.*** are rich in vitamins.

Many red algae yield gelatinous carbohydrates which are used as pudding and stabilizers. It is known as **carrageenin**. It is obtained from the cell walls of ***Chondrus crispus*** and ***Gigartina mamillosa***. The chemical is a polysaccharide complex of D-galactose-3, 6-anhydro-D-galactose and monoesterified sulphuric acid. Some other important carbohydrates are obtained from ***Bostrychia***. They are sugar alcohols like dulcitol and sorbitol.

It should be noted that the algae used as food are popularly known as **Tsao** in China, **Timu** in Tahiti and **Limu** in Hawaii.

An account of five important seaweeds is given below :

1. ***Laminaria digitata*** (Oarweed). It belongs to Phaeophyta. It is found growing in the north temperate Atlantic coasts of Europe and North America. Its young stems are widely eaten in Scotland and Ireland.

The plant has very large and broad fronds.

L. digitata has a cylindrical stem or stipe. It is 1 to 5 feet in height. The holdfast is highly branched. The blade is digitately branched.

L. digitata yields **alginates**. Other species of ***Laminaria*** are used for culinary purposes in China, Japan and some other countries.

2. ***Ascophyllum nodosumi*** (Knotted Wrack). This alga also belongs to Phaeophyta. It grows on the beaches of countries of temperate Atlantic. They are usually mixed with seaweed meals to feed the cattle because they are not very much liked by human beings.

   The plant looks like some of the species of ***Fucus***. It has branched fronds. The fronds bear variously shaped and differently sized bladders.

   The alga is also widely used for extracting alginates. They are used as stabilizer and thickener in ice creams, jellies, puddings, soups, soft-drinks etc. A preparation of sausage called **'skins'** is also obtained from alginates.

3. ***Porphyra umblicalis*** (Laver). It is belongs to Rhodophyta. It is widely used as food plants by inhabitants of south Wales, Ireland and some other European countries. It occurs widely on rocks of the coastal regions of United Kingdom, and other North Atlantic countries. ***Porphyra spp***. are found throughout the world. Laver has a large frond which is thin, and soft and has wavy margins. It has a small holdfast with which it attaches to the rock. The colour of the frond is rosy-purple, olive-green or brown. Laver is sold in the market after boiling it. In South Wales it is eaten along with potatoes and butter. It is coated with oatmeal and fried. It is then served at the breakfast with bacon and eggs. ***Porphyra spp.*** are also eaten in China and Japan. In Japan ***Porphyra spp***. are sunk along with bundles of bamboo shoots near the sea-coast. After sometime the bundles are transferred to less saline estuaries where they grow very well.

   The Chinese and Japanese use ***Porphyra spp***. in pickles, sweetmeats, soups, stews and as covering of balls of boiled rice. ***Porphyra*** is rich in vitamins B and C.

4. ***Chondrus crispus*** (Carrageenin). It is popularly known as **"Irish Moss".** It also belongs to Rhodophyta. The alga grows on temperate North Atlantic coasts of U.S.A., U.K., Ireland, etc. The alga has highly branched frond. The thallus is so much repeatedly branched that it gives the appearance of a composite fan like structure. The thallus is attached to the rock by a disc-like holdfast.

The alga gives an extract known as **carrageenin.** It is manufactured on a large scale in United States.

The carrageenin is used as an emulsifier in ice cream, table jellies, salad dressings, soups, confectioneries, etc. It is also used for treatment of coughs.

5. ***Rhodymenia palmata*** (Dulse). It also belongs to Rhodophyta. The alga has a dichtomously or palmately branched thallus. There are small thin 'leaflets' borne on the margins of the older part of the thallus. The frond is leathery and is dark red in colour.

Dulse is often used in dried seaweed. It is also eaten fresh or after cooking. Sometimes it is used for preparing alcoholic beverages. It is also used as a chewing gum.

10

Edible Mushrooms

Mushrooms have been eaten by man since ancient times. They were, however, cultivated for the first time in the seventeenth century only. Mushrooms were initially cultivated in France and thereafter in the entire Europe. Mushroom cultivation got a foothold in U.S.A. in second half of the nineteenth century. Today more than 60 species of mushrooms are cultivated worldwide inside specially built houses, green houses, cellars, etc. With greater knowledge of nutritional requirements of the mushrooms, cultivators have taken to mushroom farming on a large scale.

Mushrooms belong to the family Agaricaceae of the Basidiomycotina. They are known as "Khummi" in India. They are cultivated in Europe, U.S.A., India and many countries. The most common edible mushroom is ***Agaricus* (*Psalliota*) *campestris*** (syn. ***A bisporus***).

The cultivation of mushrooms is preceded by preparation of beds. Partially decomposed straw is mixed with horse manure and the mixture is allowed to ferment in compost heaps of 1—2 meters (in height) under controlled temperature, aeration and humidity; Once the compost is ready it is transferred to trays. The trays filled with the composted manure act as beds. Now mycelial pieces of the size of groundnuts taken from pure cultures of the fungus are introduced into the beds. The process is called **spawning** and the beds are called **spawn** or **seed**. The fungal hypha grow both outside as well as inside the compost. The beds are covered with '**vermiculite mixture**' i.e., soil or peat after a lapse of two to three weeks. The process is known as "**casing**". Casing is essential for inducing the fungus to form the fruiting bodies after a months' time. Once fruiting bodies start making their appearance the temperature is maintained at less than 15°C. Lower temperature (13° – 15°C) is ideal for the formation of the fruiting bodies. Higher temperature results in group of the mycelium only and is detrimental for the fruiting bodies. The fruiting bodies are carefully removed and packed into polythene bags for marketing purposes. They are stored in cold-storage godowns. They are cooked like vegetables and the preparation is one of the most prized delicacies available to the human palate.

Some of the more common edible mushrooms are described below:

1. ***Agaricus campestris*** (**common** or **Field Mushrooms**). It is a wild-growing mushrooms of meadows and pastures. It grows in summer and autumn. It's cap is white and rather silky expanding to about 4 inches across. The fragments of the **veil** remain attached to the edge. The stipe is rather short, not longer than the diameter of the pileus. The **gills** are initially white. They turn to rose-pink, purplish brown and then to black in colour. This mushroom has a very good flavour.
2. ***A. bisporus.*** It favours old dung. The cap is generally pale brown. The fungus differs from the previous species in having only two basidiospores per basidium instead of four.
3. ***Pleuritus ostreatus*** (**Oyster Mushroom**). The fungus grown on old tree trunks, stumps and posts, during the late autumn. It grows in clusters in shell-shaped cap. The colour of the cap is deep bluish-grey or almost blackish in the young stage but is paler, or brownish soon afterwards. The surface of the cap is smooth and moist. The edge of the cap is initially turned downwards. The stem is either absent or very small. If present the stem is attached to the cap eccentrically. The gills are white or pale yellow. The flesh is white, thick and firm. The mushroom grows in colonies and, therefore, can be easily collected.
   The mushroom lacks flavour. The fruiting bodies are stored after drying. Only young fruiting bodies are cooked.
4. ***Marasmius oreades*** (**Fairy-ring Champignan**). It is the well known fairy ring mushroom. The fairy rings are found on lawns, gardens etc. The fruiting bodies are produced in large numbers during the rainy season. The cap is brownish with a tinge of pink colour. In a dried state they look paler. The cap is initially bell-shaped and later on is flattened with a small dome in the centre. The stem is similar to the cap in colour. It is slender, tough and elastic. The gills are initially whitish. Later they get buff coloured. The flesh is whitish or coloured with a pleasant taste and mushroom like flavour. The fungus is stored well after drying. It retains its flavour well. In dried form it is very good for flavouring soups and stews. It is used for preparing ketchup and is also pickled.
5. ***Lepiota procera*** (**Parasol Mushroom**). It is a large **"toadstool".** Toadstools are larger fungi than the common mushrooms but have similar caps, gills, etc.

It grows in grassy places, often near the tree. The cap is parasol-shaped with a distinct dome in the center of the cap. The cap is 7 or 8 inches across borne an a stem of 11 inches or so. The stem is brown-mottled. The stem is tapering upwards. The cap is covered with dull brown, rather shaggy scales. The gills are white. The fungus is eaten when young and is very much liked.

6. ***Lepiota rhacodes*** **(Shaggy Parasol)**. It is sturdier that ***L. procera.*** It grows on humus-rich soil and in the woods. The cap is rounded without a central dome. The cap is 3 to 7 inches across. It is covered with large yellow brown scales. The stem is long and stout, white and smooth. The gills are white. The flesh is white but turns reddish when cut.

11

Beverage Plants and Beverages

Beverages have always found a place of pride in almost all the societies. The alcoholic beverages are derived from cereals and fruits like grapes etc., whereas the non-alcoholic beverages are prepared from a number of other plants. In the latter category the caffeinaceous drinks–coffee, tea and chocolate are very popular. They are consumed all over the world. Coffee and tea are among the foremost items of international trade. The three non-alcoholic beverages mentioned above are described below:

COFFEE

(Coffea spp.)

Origin and History

Coffea has 90 species of which four are important from the point of view of beverage. Most of the species are native of Africa and Madagascar. ***Coffea arabica*** is the only polyploid (tetraploid, $4x = 44$) species. The ***arabica*** type is indigenous to **south-western Ethiopia** in the north-eastern Africa from where it spread to Yemen in the fourteenth or fifteenth century. From Yemen it first spread to the Malabar coast of India and then to Sri Lanka and Java (Indonesia).

It was introduced in Arabia only 500 years ago. Pilgrims and traders carried it via Turkey to Europe in the sixteenth and seventeenth centures (Venice 1615; Paris 1645; and London 1650). The plant was introduced in Sri Lanka, India and Java by 1700, in the West Indies in 1720 and in Brazil in 1770. It was carried to U.S.A. in the nineteenth century.

Canefora (robusta) coffee, a diploid species, has its centre of origin in **central and western equatorial Africa** and in **Madagascar** (Fig.). It was introduced (20 plants) into southeast Asia (Indonesia) about 100 years ago from Zaire. ***Canefora*** coffee was found there to be a very good substitute for the *arabica* type which was earlier devastated completely by the coffee leaf rust disease (causal organism = ***Hemileia vastatrix***).

U.S.A. is the greatest coffee-consuming country. The world production of coffee is more than 5 million tons, out of which South America, particularly Brazil produces 4 million tons of ***arabica*** type. ***Arabica*** type constitutes 80 per cent of the total production of coffee.

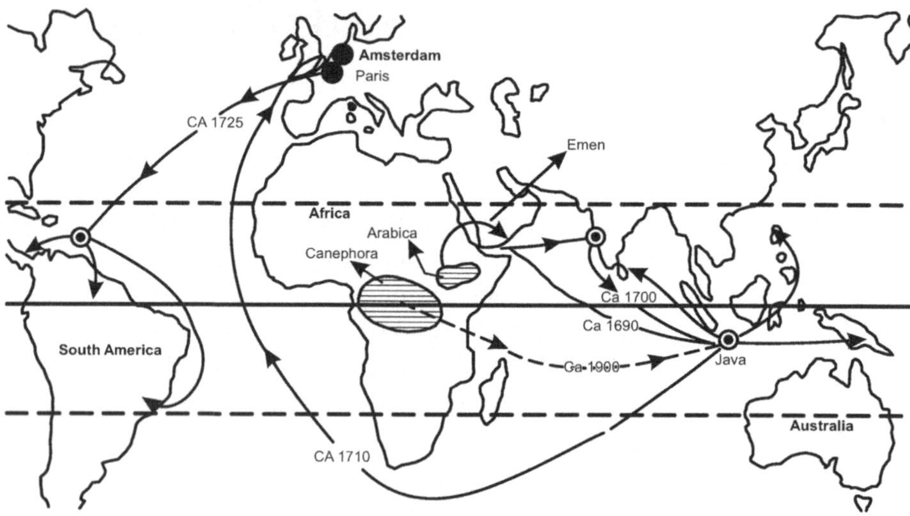

Fig. 11.1. Distribution of the cultivated, ***Coffea arabica*** and ***C. canephora***.

Botanical Description

Coffee plant belongs to the family **Rubiaceae**. The are three important species of coffee plants. They are:

(i) ***C. arabica*** (Arabian coffee).

It is the source of best quality coffee and accounts for 90% of the total world production of coffee,

(ii) ***C. robusta*** or ***C. canephora*** (Congo coffee),

(iii) ***C. liberica*** (Liberian coffee).

The coffee plant is a shrub or small tree. It has a height of 15 to 50 ft. It bears opposite leaves which are dark green. The plant bears white axillary flowers two or three times a year. The fruits are two-seeded or occasionally one-seeded drupes (Figs. 11.1, 11.2 and 11.3). The coffee 'berries' have fleshy mesocarp and thin membranous endocarp. The leaves and the fruits contain caffein (I to 2%). The green seed contains 34% cellulose, 8% chlorogenic acid, 10 to 13% oil, 7% sugar, 14% protein, 12% water and some other components.

Cultivation

The plant grows better on highlands of a height of 4,500 ft. or so. On the other hand they can grow on plains as well as on high altitudes upto 6,000 ft. They like moderate temperature and frequent mists of the hills (moist climate). They grow best on fertile, well drained soils.

The seeds are sown first in seed beds. The seedlings are transplanted in rows about 9 ft. apart. The plant bears fruits from third year and continues to do for about 25-30 years. Planting, pruning and picking of the berries are all done by hand. Picking is done when the fruits are fully ripe. They may be stripped off or be allowed to fall down on the ground. They are then winnowed by throwing them in air.

Preparation

There are two methods for the preparation of coffee—**dry** method and **wet** method. In the dry method the coffee berries are dried in the open. The dried skin and pulp are separated from the seeds by mechanical means, or by pounding them in a mortar.

In the wet method the fruits are taken to the floatation tanks where the ripe berries float and the debris including the green berries sink. The seeds are removed from the pulp by means' of pulping machines. Parts of the pulp sticking to the seeds are removed by controlled fermentation over a 12 to 24-hour period. Seeds are then thoroughly washed in water and are sprayed on salts in the open for drying. The brittle, dry husk like parchment (endocarp) is removed by machines and the silver skin is removed by polishing.

Fig. 11.2. Coffee-flowers and fruits (Courtesy: Pan America Coffee. Bureau, New York).

The greenish-grey seeds are graded and are then usually packed for export purposes. The coffee 'beans' are roasted at 500°F for 5 minutes. The roasting changes the colour of seeds from green to brown. The roasting results in the development of certain typical aromas (essential oils and caffeol etc.) that give the coffee its appetizing flavour. The aromas are due to volatiles like carbonyls *viz.,* acetaldehyde, isovaleraldehyde and acetone, alcohols, ketones, furfuryl compounds, amino acids, phenols, etc.

The water soluble constituents increase to 35 per cent. The chlorogenic acid is partly hydrolysed into caffeic and quinic acids. These three acids along with trigonelline are responsible for the bitter and astringent taste of the coffee brew. Organic acids like tartaric, citric and malic cause an acidic taste.

Fig. 11.3. Blending of coffee being done by an expert coffee taster (By courtesy of Pan American Coffee Bureau, New York).

Grinding of the roasted 'beans' is the last step in the preparation of coffee powder. 'Instant' coffee is prepared by concentrating a strong extract of coffee powder under vacuum. Synthetic flavours are often used to make up the loss of aroma components during the process.

Roasting and grinding are done only few days before coffee is to be consumed otherwise it loses the flavour and the oils become rancid. The coffee from different sources are tasted by a 'coffee taster' for blending them into a particular kind. The roasted coffee beans have 0.5 to 1.5% caffein, an alkaloid.

Caffeine

The commercial coffee contains 0.8% to 1.8% caffeine. Instant coffee contains 4.4% caffeine. One cup of coffee roughly contains 100 milligrams of caffeine. The solubility of caffeine in water is only 2.2%. The caffeine after assimilation enters the blood stream and reaches all parts of the body within five minutes. It is, therefore, a potent drug, which is both beneficial and harmful depending upon the dose and frequency of intake. The amount of caffeine which gets distributed in various tissues is approximately in proportion to their water contents. In the tissues the caffeine is quickly decomposed into 1,7-

dimethylxanthine, i.e., paraxanthine. The demethylated products appear in the blood. It, however takes about one week for the complete clearance of caffeine through urine. Caffeine acts on the central nervous system. Sensory stimuli and motor activity act at a quick rate. It prevents drowsiness and fatigue. It relaxes the muscles of the bronchial tubes. It stimulates medullary, respiratory vasomotor, vagal centres and spinal cord with increased reflex activity and excitation of lower motor centers. It has an over-all effect on the cardiovascular system. It casuses the constriction of the muscles of the heart and dilates the coronary, pulmonary and general systemic blood vessels resulting in increase in the rate of flow of blood to every part of the body including brain.

Hazards of Caffeine

There are more than 10,000 scientific reports about the undesirable effects of caffeine. The alkaloid has been found to be responsible for several health hazards. The addiction to coffee is such that stoppage of the drink results in headache, irritability and sometimes nausea. Luckily addiction to caffeine can be got rid off in a few days. Excess of caffeine can turn a person into a nervous wreck. The "restless legs syndrome" of psychiatric patients has now been partly attributed to the excess intake of caffeine. Excess of caffeine causes over-excitement, insomnia, mild delirium, sensory disturbances such as ringing in the ears and flashes of light, restlessness, fluttering of heart, heart diseases, peptic ulcer, tumours and birth defects.

Caffeine is more hazardous to children since they cannot eliminate it as fast as an adult can. The caffeine containing cola drinks are a great hazard for the children. The cola containing cold drinks contain 35 to 55 milligrams per (360 ml) bottle. Administration of very high concentration of caffeine during pregnancy has been shown to cause foetal deaths and malformation in mice. It is therefore, advisable to avoid these hot and cold drinks during pregnancy particularly in the early stages. The soft drink industry should remove caffeine from drinks for the safety of children. It should be noted that the morning cup of coffee or tea gives the feeling of well being with greater concentration and intellectual activity followed by, sadly enough, decreased efficiency later during the day.

Uses of Coffee

1. The presence of a stimulative alkaloid in coffee makes it a much wanted beverage. It relieves fatigue and stimulates the physical and mental activities.
2. In Sumatra coffee leaves are used like tea leaves to prepare beverage.
3. Caffeine is used in several drugs. It relieves headache by its vaso constrictor effect on blood vessels.

4. It is used in the treatment of cases of poisoning caused by alcohol and morphine.
5. It is given along with ergot to relieve pain of migraine.
6. It has therapeutic value in the treatment of bronchial asthma as it relaxes smooth muscles of bronchi.
7. It is diuretic i.e., causes increased production of urine.
8. The waste products, pulp and parchment are used as fertilizer and fuel. They are also used for the manufacture of a plastic material, cafelite, used for insulating purposes.

TEA

(Camellia sinensis)

Origin and History

Tea is the most popular beverage. It is believed to be indigenous to **southwestern China, northeastern India (Assam)** and the adjoining areas of **Upper Burma**. Tea is found in a wild state in these areas. Tea has been cultivated in China since early times. Tea is believed to have spread from its primary centre of origin, lower Tibetan mountains or central Asia to a secondary centre, the sources of Irrawady. From the latter site tea spread in three directions giving rise to three types of tea *viz.,* China, Assam and Cambodia. It is assumed that the small-leaved China tea gave rise to large-leaved Assam tea. Southern Assam is believed to be a secondary centre of origin. **Wight** (1962) has named the two important types of tea plants as ***C. sinensis*** (Chinese) and ***C. assamica*** (Assamese). The Cambodian type is classified as a subspecies of ***C. assamica.*** Tea was carried to Japan in 1000 AD. The Dutch introduced it into Europe in 1610 AD. It reached London in 1664 and Boston (U.S.A) in 1714. Tea is now widely cultivated in the tropical and temperate regions extending from 40°N in Russian Transcaucasia to 33°S in Argentina.

Tea Growing Regions

The chief tea-growing regions are all situated in Asia, *viz,,* China, Japan, Formosa (Taiwan), Java, Sumatra, Bangla Desh, India and Sri Lanka. Commercial tea plantations were started in India, Sri Lanka and Indonesia in the nineteenth century. China, India and Sri Lanka are the world's largest three tea producers. India cousumed 800 mkg tea in 2006-2007. China produces 1000 million kilograms. In 2006-2007 India produced about 950 million killograms.

In India tea is cultivated in Assam, northern districts of Bengal, Kerala and the Nilgiri hills. Some other tea-growing areas are Ranchi, Dehradun, Kangra and Kumaon districts.

Botanical Description

Tea belongs to the family **Theaceae**. The plant is an evergreen tree or shrub which can grow to a height of 30 to 50 ft. It is, however, not allowed to grow to that height (Fig 11.4). Under cultivation it is prunned down to a small height of 2 to 5 ft so that it looks more like a shrub. It bears alternately arranged

Fig. 11.4. A photograph of tea garden.

leaves which are elliptical in shape and serrated in margin. They are glabrous. The young leaves are however, pubescent. The leaves contain numerous oil glands. Old leaves are bright green coloured. They may be up to 12 inches in length. White and fragrant axillary flowers may occur singly or in clusters of few (Fig. 11.5). The fruit is a 1 to 4-celled capsule, each chamber containing 1 to 3 seeds. Tea has about 1,000 varieties.

Fig. 11.5. Flowers of tea (Courtesy: L.S. Cobley).

It grows well in tropical and warm temperate climates having plenty of rainfall. A minimum annual rainfall of 60 inches is needed. The rainfall should be evenly distributed throughout the year. Whereas the Assam and Bengal plantations get plenty of rainfall 50"–150" and 100"–200" respectively the tea farms of Ranchi and Dehradun are poor in growth due to uneven distribution of rainfall. Prolonged drought damages the tea crop.

Ecological Factors

High temperature accompanied with rainfall is good for tea plant. The optimum monthly temperature between 70°–90°F is excellent. Temperature below 65°F is not good for its growth. Frost damages the crop. Shady conditions are better than exposed situations.

Though tea has been cultivated on hilly tracts up to an altitude of 7,000 ft. it is usually grown in areas lying between 3,000 and 4,000 ft. above the sea level. It can, however, be successfully grown on flat plains provided the soil is well drained. Some of the finest tea plantations in Assam are on levelled grounds, lying between an altitude of 50–400 ft.

Well manured and humus rich deep soils are good for the plant though it can grow on poor soils as well. The soil usually ranges from the light sandy to stiff clayey types. Acidic soil (pH 5.2 to 5.6) is good for growth to the tea plant. The soil must be well drained.

Cultivation

Nursery. Tea is usually grown by sowing seeds. Carefully selected seeds are sown in well prepared nursery beds. The nursery beds are usually shaded. Seeds are sown 4"–8" apart. The nursery is manured till the seedlings are 6–9 months old.

Preparation of land. The land where the tea seedlings are to be planted is also prepared. Land previously occupied by forest is best for tea plantation. It is first completely cleared of trees and is then ploughed to remove the roots etc. The steep slopes of the hills are contour terraced. Drains are usually dug along the contour of the slope to prevent erosion of soils and waterlogging of the low-lying areas. The size and number of drains vary with the steepness of the land, the nature of the soil and the amount of rainfall.

Planting of seeds. The seedlings are planted in holes dug 3 ft. to $4\frac{1}{2}$ ft. apart depending upon whether land is terraced or is flat. Organic manure is mixed with the soil for filling the holes. The plantations are shaded by growing leguminous plants spaced about 50 ft apart.

Weeding. There are a number of methods to keep the tea plantations free from weeds. The weeds are prevented from growing by scraping the soil surface and also by deep cultivation.

Pruning. Pruning is done for three purposes: *(i)* To induce vigorous growth of the tea plant for development of new shoots; *(ii)* To ensure a continuous supply of tea leaves to the tea industry and *(iii)* To keep the height of tea plant reasonably low so that the leaves can be plucked by hand.

Pruning is usually of two types– *(i)* light pruning and *(ii)* heavy or medium pruning. When the plants are 2-5 years old, during their slow growing period, the first few prunings are done. The light pruning may be done annually or biennially. Heavy pruning is resorted to after ten years or so when the plant is cut back to the ground level so that the suckers replace the old bush.

Plucking. Plucking is started when the plant is 4 to 5 years old. The quality of tea depends upon the age of the leaves. The tannin content of the leaves is variable with their age. The young leaves with more tannin make better tea than old leaves with less of tannin. The percentage of tannin in the tea flushes is as follows: bud–25, 1st leaf–28, 2nd leaf–21, 3rd leaf–14, stalk between 2nd leaf and bud–12, and stalk between 2nd and 4th leaf–6. It is obvious from above that picking of the terminal bud only or with 1st leaf gives excellent quality of tea. The bud is usually plucked along with 1st and 2nd leaves. Inclusion of more than two leaves gives 'coarse' tea. Sometimes all the 4 or 5 leaves of a 'flush' are collected separately to prepare 4 or 5 different qualities of tea.

Plucking is normally done by hand by women and children. Plucking may be done at intervals of 7–10 days since the new shoot develops very quickly. In cold climates there may not be any plucking during winter months because the growth of the plants almost comes to a stop. In hotter regions 25 or 30 pluckings can be made in a year.

Though there may be individual bushes of tea surviving for as many as 200 years, the yield decreases substantially after the plant is 40 to 50 years old.

Preparation of Tea

Commercially 4 kinds of tea are prepared: **black** tea, **green** tea, **oolong** tea and **let-pet** or **leppet** tea.

Black Tea

Black tea is by far the most important of all. There are four steps in the preparation of black tea.

Withering. Harvested fresh leaves are sent to the factories where they are withered for about 24 hours on well ventilated indoor shelves or racks. The leaves may be exposed to sun or warm air may be allowed to pass over them in the houses which have open sides. The withering results in decreasing of the water content of the leaves which become soft and flaccid. Withering is usually omitted in the case of green tea preparation.

Rolling. Withered leaves are sent to the rolling machines, which in several rolling and breaking operations damage the individual leaf cells so that the juice and enzyme contents are released and the whole rolled mass gets smeared with them. This facilitates **fermentation**. The polyphenols called **catechins** (25%) are oxidised into **theaflavin** and **thearubigins** by the enzyme **polyphenol oxidase**. They give the characteristic taste, colour and strength to the brewed tea.

Fermentation. Rolled mass of curled tea leaves are spread out in cool, moist fermentation rooms. They are usually covered to keep them warm (75°–60°F). Very high temperature spoils the quality of tea because of over-fermentation. The leaves are allowed to ferment for 2–6 hours, during which period they change themselves from green to bright reddish coppery colour and develop pleasing aroma. Fermentation is not needed in the preparation of green tea.

Drying or Firing. The fermented leaves are conveyed through air chambers at 130°F or so for 30 to 40 minutes and at 192–200°F for a while in the end. The leaves coming out are left with only 3 to 4% moisture.

Grading. The tea prepared in the manner mentioned above is graded into various kinds. The leaf grades (orange pekoe, pekoe) give better tea than 'broken' grades (broken orange pekoe, broken pekoe etc.). The broken grades are used for quick preparation of tea particularly in restaurants, canteens etc. Broken grades can be packed in larger amounts in a container as compared to leaf grades.

Packing. The tea is packed in plywood boxes (24" × 19" × 19")' which are lined with aluminium foil or parchment paper to protect them from moisture. The small tea packets are lined with metal foil.

Blending. Blending is an art by which the different qualities of a particular tea brand produced in a tea estate are blended together to give a uniform product.

Green Tea

As slated above the preparation of green tea does not involve the processes of withering and fermentation.

Its preparation includes heating, rolling and drying. The latter two processes are similar to those of black tea preparation. Heating may be done either by steaming or by pan firing. The heating is completed within a few minutes so that the green leaves are without any sign of scorching or reddening.

Oolong Tea

An exclusive produce of Taiwan (Formosa), oolong tea is a favourite of Americans. The manufacturing process is similar to that of black tea except that the oolong tea is allowed only a light fermentation. The semi-fermented tea has a characteristic flavour more because of the particular variety of tea grown under a particular set of soil and climatic conditions of Formosa and not because of any marked difference in its manufacturing process. A similar semi-fermented tea is the scented jasmin tea. It is prepared by allowing the fragrant jasmin flowers to dry along with the tea leaves. The flowers are later on removed.

Brick Tea

It is usually manufactured in China from where it is exported to Tibet and Russia' Brick tea is prepared from coarser leaves, twigs and remains left over from the preparation of black tea *viz.* dust, stalks, etc. They are steamed over a boiler. They are then usually mixed with a little rice paste before they are pressed into moulds to give firmness.

Let-Pet or Leppet Tea

In Burma boiled or steamed green leaves of tea are preserved in pits. They are used more as pickles or vegetables than as beverages.

Uses of Tea

Tea contains a caffein like alkaloid called **theine** (2 to 5%), a volatile oil, and 14 to 28% tannin. While tannin does not add to the drinking quality, the volatile oil and the alkaloid theine which dissolve quickly in hot water give a refreshing and stimulating beverage. It relieves the body fatigue.

Excessive drinking of tea is considered to be harmful for the digestive system.

COCOA AND CHOCOLATE

CACAO

(Theobroma cacao)

Origin and History

Cocoa and chocolate are the most nutritious of all beverages. They are derived from the seeds of the cacao tree, a native of South America. It was a favourite beverage of the Aztecs, who made a cold, frothy liquid seasoned with peppers and spices and called it 'chocolate'. Spaniards introduced it into Europe in the sixteenth country, when along with coffee it became a very popular beverage.

Initially it was grown in Brazil, Equador, and in other neighbouring regions but today it is also grown extensively in Africa (Ghanna, Nigeria, etc.), Java and Sri Lanka.

Two-thirds of the world production of cacao is from Africa (Ghana and Nigeria) while the rest is obtained from South and Central America and the West Indies. Indonesia and Sri Lanka produce a very small fraction of the crop.

The crop after processing is mostly exported to the temperate countries which are the leading consumers. Europe consumes more than 50 per cent and America about 40 per cent of the world's production of cacao.

Botanical Description

The cocoa tree belongs to the family **Sterculiaceae**. It is a branched erect tree. It is small in height (15 to 25 ft). It bears long leaves which may be up to 12" in length. The trunk and the larger branches of the tree bear flowers and fruits on short stalks (Fig. 11.5). The pod like fruits bear 49 to 60 large seeds known as 'beans'. The fruits start developing in 4 to 5 years old trees. There is no particular flowering season and the plant produces flowers throughout the year. Several crops of fruits are, therefore, harvested in a year.

A

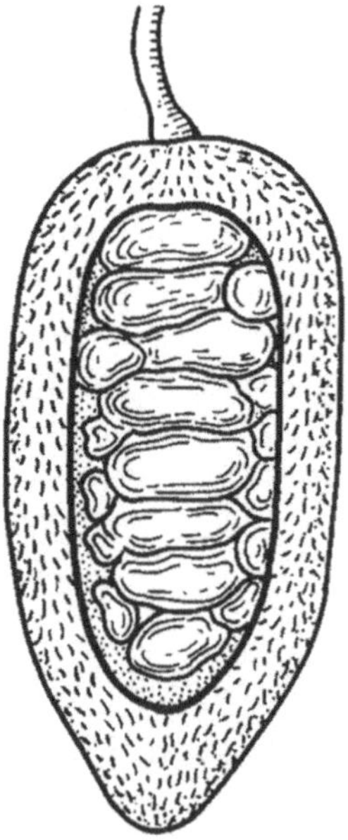

B

Fig. 11.5. The pod-like fruits of the cacao tree are borne directly on the trunks. In the middle left is a flower and on right a section of the fruit showing seeds.

Ecological Factors and Cultivation

It requires a hot and moist climate of the tropics and, therefore, it grows well within 20° of the equator. It, however, needs protection from direct rays of sun and also from strong winds. This is why cocoa is grown along with cash crops and permanent shade trees. Well drained, moist, deep alluvial soil is necessary for its cultivation. It is sensitive to low temperature and high attitudes. It cannot grow below 60°F and at altitudes beyond 2,500 ft.

The seeds are sown in prepared seed beds. The seedlings are transplanted 4 ft apart.

Preparation

The pods are cut with knife and the pulp of the fruit and the seeds are taken out. They are dried in the sun or are allowed to ferment sufficiently. The seeds are washed to remove the disintegrating pulp. They are then dried and polished. The beans are sorted and roasted at high temperatures in iron drums. Roasting results in increase in the fat (50%) and protein contents, decrease in tannin content and in the development of characteristic aroma, as a result of complex chemical reactions occurring in the cotyledons.

The seeds are broken into pieces called 'nibs'. The nibs are made into an oily paste called 'liquor'. Removal of oily cocoa butter and making powder of the residue results in cocoa. If the cocoa butter is left in and sugar is added chocolate is formed.

Uses of Cocoa and Chocolate

1. They are an excellent beverage. The high fat content of the bean is reduced by half (to 25%). The rest of the extracted fat is sold as cocoa butter.
2. They contain considerable nutrient material in the form of fats, proteins and carbohydrates and, therefore, they are nutritious.
3. Cocoa butter is a very fine edible fat. It is used for a number of preparations.
4. It is used in the finest of chocolates, candies and cakes etc.
5. The 'cake' gives caffein like alkaloid theobromine which is used in soft drinks "colas" and for other purposes.
6. The cocoa shells are used as a cattle feed and as fertilizer. They can also be used as a beverage and as an adulterant of cocoa and chocolate.

Other Caffein Containing Beverages

Mate. It is also called Paraguay tea. It is obtained from the leaves of various species of holly, chiefly ***Ilex paraguariensis***.

Gaurana. It is prepared from the seeds of ***Paullinia cupana*** of the Amazon valley.

Khat. It is obtained from the leaves of ***Catha edulis*** of north-eastern Africa.

Cola. This is obtained from the seeds of ***Cola nitida***. The seeds are usually called cola or kola nuts.

Cassine. It is a tea like beverage. It is obtained from ***Ilex vomitoria*** of Mexico.

Yoco. It is obtained from the bark of **yoco** ***(Poullinia yoco)*** of Peru, Equador and Southern Colombia.

Other Beverages

Other non-alcoholic beverages are normally called ***soft drinks***. They are usually ***juices of*** fruits like orange, lemon, apple, pineapple strawberry raspberry, etc. The fruit juice is pasteurized and is mixed with preservative before bottling or canning it. The fruit juices are rich in sugars and are, therefore, good sources of energy. These days synthetic flavours are increasingly used in commercial products.

Other soft drinks are **malt beverages**, **ginger ale**, **sarsaparilla**, **col beverages** and **soda water**.

Alcoholic Beverages

Alcoholic beverages are consumed worldwide by rich and poor in developing countries as well as in developed countries. Alcoholic beverages are classified into two categories

(i) Fermented beverages and *(ii)* Distilled beverages

FERMENTED BEVERAGES

Wine

Grapes (*Vitis vinifera*) are the chief source of wine. Wine contains 10 to 20% alcohol. The main wine producing country is France. **Bordeaux, Burgundy** and **Champagne** are the best varieties of wine prepared in France. The other varieties of wines produced in other countries are Hungary—**Tokay;** Germany—**Rhine;** Italy-**Chianti** and **Asti**. Portugal produces **Port** and Spain produces **sherry**.

The first step in the preparation of wine is the expression of juice from grapes and fruits like berries.

The yeasts are introduced into the chamber for carrying on alcoholic fermentation. After some lapse of time the medium is heated to destroy the yeasts so that further fermentation does not take place. For preparing good quality wine the wine is allowed to ripen in absolutely airtight vessels for about 5 years and even for decades.

Wines are classified on the basis of several criteria—the percentage of alcohol, colour, quality of grapes etc.

(i) Sweet wine	—	Incompletely fermented sugar
(ii) Dry wine	—	Completely fermented sugar
(iii) White wine	—	From white grapes
(iv) Red wine	—	From red grapes
(v) Sparkling wine	—	CO_2 rich wine
(vi) Fortified wine	—	Wine has a stronger content of alcohol due to addition of brandy etc.

Beer

The chief sources of beer are the cereals particularly rice, maize, barley, rye, etc. It contains 3 to 6% alcohol.

Beer is manufactured in two steps: *(i)* Malting and *(ii)* Brewing.

(i) **Malting.** The enzyme diastase present in the germinating seeds converts the starch of the grains into sugar. The grains are kept in water for 1 to 4 days. When the grains have germinated they are dried for about 12 hours.

(ii) **Brewing.** The dried malt is crushed and is then heated to 170°F in a medium of water. The expressed sugar content is boiled for 2 hours and is then cooled rapidly. At this stage yeast cells are added to the medium containing the malt for fermenting the sugars. The CO_2 produced during fermentation is not allowed to escape by putting the liquid inside airtight bottles. The fermentation process is stopped midway by adding hops (***Humulus lupulus***). The plant also imparts characteristic flavour and bitterness to the beer. The foaming of the beer when poured in a glass is due to accumulated carbon dioxide.

Root beer is prepared from the extracts of roots, barks etc. of ginger, wintergreen in which sugar and yeast cells are added.

Hard Cider

Hard cider is prepared by fermenting apple juice. Apple juice can also be fermented to obtain acetic acid (vinegar).

DISTILLED BEVERAGES

Whisky

Whisky is a very popular alcoholic beverage and is diluted with water or soda-water. It has a very high percentage of alcohol upto 50%. The principal sources are the cereal grains of wheat, barley, maize, oat, rye. Potato is also used for preparing whisky. Their fermented mash of malted or unmalted are distilled to obtain "low wines", which is further distilled to produce "high wines".

Whisky is prepared by mixing water with "high wines". The whisky is aged to get rid of the bitter volatile components. The aging of whisky is carried on for at least 4 years or more within charred oak containers to obtain the finest quality. **Scotch whisky** is always prepared from barley malt and **Vodka,** the Russian whisky from fermented wheat mash.

Brandy

Brandy is obtained by distilling wine. It has 60 to 70% alcohol.

Rum

It is obtained by distilled unrefined products of sugar cane like molasses etc. It has 40 per cent alcohol.

Gin

It is distilled from the ferment mash of malt or raw grains of barley and rye. It is flavoured with an extract of Juniper berry or other plant products. It has a low content of alcohol.

QUESTIONS

1. Name three beverage yielding plants and give their botanical names. Describe the origin, distribution and method of preparation of the plant product.
2. Describe the origin, distribution, botanical description and cultivation of coffee yielding plant.
3. Name any industry dependent on plant material. How is the latter processed and what are its uses?
4. Describe the morphology, method of cultivation and processing of a beverage which earns foreign exchange.
5. Write a short account of the processing of tea leaves. What do you mean by black tea and green tea?

12

Fibers

The fiber plants of the world are second only to the food plants in importance. Man has been dependent on these plants for his clothing and for a variety of other needs from times immemorial. The important fiber plants were cultivated during the ancient civilizations. The commercially important fiber plants are, however, not many.

Botanically many vegetable fibers are sclerenchyma cells which are thick walled, slender and tapering with pits. Chemically the fiber cell walls are composed of cullulose only or of cellulose, lignins, hemicelluloses, etc. Commercially the term fiber includes hairy or thread-structures borne by certain plant organs.

Fibers are often classified according to their utilization: *(i)* Textile fibers, *(ii)* Brush fibers, *(iii)* Plaiting and rough weaving fibers, *(iv)* Natural fibers, and *(v)* Paper-making fibers.

Fibers excluding the wood fibers are also classified into the following three types depending upon their origin and structure:

(i) Soft, stem or bast fibers,

(ii) Hard, leaf or structural fibers,

(iii) Surface fibers.

Soft fibers. They are present in groups in the cortex of the dicotyledonous stems and are usually associated with the **phloem *(bast)*** or the **pericycle**. They are sclerenchymatous with deposits of lignin in the cell wall. The fiber or clusters of fibers are separated from the rest of the stem by retting in water. The examples of such fibers are jute, flax, ramie and hemp, etc.

Surface fibers. They are produced on the, surface of stems, leaves, seeds, fruits, etc. Cotton, the most important fiber of the world, is produced on the surface of seeds whereas the stuffing material kapok, used in the manufacture of lifebelts, is produced by the ovary or fruit wall on the inner side for the protection of developing seeds. The husk of the coconut fruits yield the coir fiber.

Hard or leaf fibers. They are found in the leaves of monocotyledonous plant's supportive and conductive strands. They are termed fibrovascular bundles as they also contain xylem and phloem. The cells are very much lignified. The whole fibrovascular bundle serves a single unit. They are separated by mechanical scrapping. They form the cordage fibers of commerce and are used for the production of ship's cables, ropes and coarse sacking. Manila hemp, New Zealand hemp, Sisal, Mauritius hemp, etc. belong to this category.

Some important fiber plants are described in this chapter.

Soft or Bast Fibers

JUTE

(Corchorus spp.)

Origin and History

Jute is the best of the coarser bast fibres and one of the most important commercial fiber crops of the world, particularly of India. The fiber is obtained from the stems of two species of an Asiatic genus ***Corchorus*** which belongs to **Tiliaceae**, a family closely related to **Malvaceae**.

The genus ***Corchorus*** includes about 40 species. They are distributed throughout the tropical regions of Africa, America, Australia, China, Formosa, Bangla Desh, Sri Lanka, Nepal, India, Java, Japan and the Malayan Peninsula etc. ***Corchorus olitorius*** is believed to be of **African** origin and ***C. capsularis*** of **Chinese** or most likely of **Indo-Burmese** origin. The former is known to grow in Africa and West Asia for a very long time.

Both ***olitorius*** and ***capsularis*** have been domesticated in India in recent times. While ***olitorius*** has the finer fiber, ***capsularis*** is more adaptable particularly to flooded conditions.

Jute is mostly confined to the Indian sub-continent. It was a monopoly crop for India till 1947 when most of the areas producing the finer quality of jute went to East Pakistan (Now Bangla Desh). Raw jute and jute products are one of the main sources of foreign exchange earning for both the countries. West Bengal and East Bengal together produce nearly four-fifths of the world's total jute output. Its cultivation is being extended to some other parts of our country. In India only 8 species are found out of which only two, ***C. capsularis*** and ***C. olitorius*** are of commericial importance. The other species are found wild in nature.

C. capsularis is believed to be native of **South China** where it is found in the wild state. The plant might have entered India from there. This species accounts for nearly 75% of the total acreage under jute because of its ability to grow on both high and low lands and in water logged conditions. ***C. olitorius***, on the other hand, can grow on highlands and cannot thrive under water logged conditions. The former species has about 50 and the latter has 8 races.

Jute is second only to cotton in production of all the natural vegetable fibers.

Jute fibers consist of about 65% cellulose, hemicellulose 25%, lignin 12% and small amounts of waxes, gums, minerals etc.

Jute fibers are not very strong. They also lack the elasticity as seen in many other fibers. They are coarse and rough. Jute fibers cannot survive for long if kept moist. They get easily spoiled in water.

C. capsularis has a bitter glucoside – **Corchorin.**

The fibers of ***C. olitorius*** are stronger, softer and far more lustrous than those of ***C. capsularis***.

The fibers are arranged in concentric rings in the form of bundles just outside the xylem, and within the soft tissues of the secondary phloem.

C. capsularis constitute three–fourth of the total production of jute in India. It grows very well in upland and lowland areas.

C. olitorins on the other hand is only a highland crop because it is very susceptible to water–logged conditions.

Botanical Description

The plant is a slender annual. It is quite tall and attains a height of 10 feet (Fig. 12.1). Stem may be branched near the tip, but when planted close together they remain unbranched, It bears alternate, simple leaves. The plant bears small groups of two or three yellowish coloured flowers. ***C. capsularis*** bears small globular capsular fruits while ***C. olitorius*** bears long cylindrical capsules.

Ecological Factors

The jute crop requires a warm and humid climate. It is cultivated during the rainy season when the temperature (60° F to 100° F) and humidity are quite high. The crop is often water logged but the young plants cannot withstand waterlogging.

Jute is cultivated on the alluvial soils in the delta regions of rivers like Ganges and Brahmputra. The soils are of three kinds: light coloured calcareous soils of fine texture, greyish soils rich in potash and red coloured alluvial soil.

Cultivation

The seed bed is prepared by ploughing and cross-ploughing the land about six times. ***C. capsularis*** can be sown at any lime after the middle of February and with earlier sowings better crops are obtained. ***C. olitorius*** is not normally sown before middle of April.

Soils with fresh silt deposition do not require manuring. Other soils are usually manured with cowdung and wood ashes. Artificial manures are not used.

The seeds are sown broadcast. During the first two months of the crop, a large number of plants have to be thinned out gradually, so that the plants become 4–6 inches apart. Weeding is done with 'khurpi' two or three times during the season.

Fig. 12.1. A crop of Jute (***C. olitorius***).

Harvesting

Jute plants can be harvested when they are 3 to 4 months old. Jute plants may be cut at any time before they are dead ripe, but harvesting is usually done at the flowering stage. The ideal stage of harvest is when the plants are in small pods. At this stage maximum yield of good quality fiber is obtained. If harvested later the quality of the fiber is reduced because they become more lignified and woody to give coarser quality. The plants are cut close to the ground with a sickle. In high lands the harvested plants are cut and left in groups at different places in the field for 24 days when most of the leaves dry up. The plants are then tied in bundles of about 6–9 inches in diameter. At the time of bundling, the plants are shaken, when most of the leaves are shed on the ground. In many places the harvested plants are immediately tied into bundles which are then laid on the ground. In 2–4 days the leaves are shed and the bundles are then taken for steeping in water.

Fiber Extraction

Retting. Retting is a process by which the fibers in the bark get loosened and separated from the woody stalk due to the removal of pectins, gums and other mucilaginous substances. This is usually effected by the combined activity of water and microorganisms. The tied bundles of jute stems are taken to the nearest pond or ditch for retting. The retting process is completed in 8–30 days. Retting can also be done by chemical processes. Retting is completed when bark separates out from the wood.

Extraction. Fibers are extracted from the stalks of the retted jute by hand, jerks and pulls. The stems are often whipped on the surface of water. Machines have also been designed and used in some countries for extracting fibers. The loosened i.e., the retted fiber are pulled apart from the sticks by hand. They are washed in water which is then squeezed out. The fibers are spread out in the sun on clean ground for 2 to 3 days to dry them. The dried fibers are then tied together in small bundles.

Fiber Character

The jute fiber is developed in the outer portion of the bast (secondary phloem) of the stem. The fiber strand consists of a loose network of many smaller strands, each of which contains many elongated cells. The cells have pointed to tapering ends and are polygonal in cross-section. Cells i.e., the individual fibres vary from 2 to 5 millimeters in length and have wide lumen. The retted strands of jute are 5 to 10 feet in length. They are weaker than flax and hemp. They deteriorate rapidly in water.

The quality of jute fiber is determined by its fineness, colour, lustre, length and strength.

The two species of ***Corchorus*** differ much in the quality of the fiber that they yield. Best fiber is obtained from ***C. capsularis***. The fiber of ***C. olitorius*** has a yellowish, reddish or greyish colour depending upon the nature of the retting water. The fiber of ***C. capsularis*** is whitish and is called '**white jute**' by the trade. The tannin present in the plant tissues combines with the iron of retting water and stains the fibre. There are a number of well-defined varieties based on various characters: pigmentation, pod shape, colour of petals and anthers and yield and quality of fiber.

Uses of Jute

1. Jute is a cheap and readily spun fiber. Jute fiber is the world's principal material for manufacturing coarse textile for sacs, bags and convases. The poorer fabrics are used in the making of gunny bags while better ones are used for the manufacture of curtains, roofing fabrics and covers for cotton bales. The bags are used in the packing of sugar, potato, onions,

pulses, coffee, cocoa, etc. It is used in the manufacture of twine, upholstery, rugs, carpets, etc.

2. The tender shoots are used as vegetable in Egypt, Sudan and other countries.
3. The jute butts (pieces of lower ends of stalks and short fibers) Are used in the manufacture of paper.
4. The jute stems, after extraction of fiber, are used for temporary fencing.
5. They are also used in the manufacture of a kind of thick paper board.
6. Jute seeds are rich in oil. 14.7% oil is extractable by petroleum ether. It can be used in the manufacture of soap and can also be converted into an edible oil. The oil from the seeds of ***C. olitorius*** is separated into non-drying and drying portions the latter of which can be used in paints and varnishes.

FLAX

(Vern. Alsi, Tisi)

(Linum usitatissimum)

The flax fiber is obtained from the fibrous bundles of the stem of ***L. usitatissimum***, which belongs to **Linaceae**. It is also a source of linseed oil (Chapter.). The plant was well known to the primitive civilizations of Swiss Lake Dwellers, Hebrews, Egyptian, Greeks, and others dating back to well over 4,000 years. The Egyptians produced finest quality of linen and used it to cover their mummies. Linseed oil was used in embalming. The early Greeks and Romans cultivated the plant for both the fiber and the seed. Ancient Indians used linseed oil in rituals.

According to **Vavilov,** the small seeded types of linseed originated in **South-Western Asia (Afghanistan)**. It spread northward to Europe and other parts of Asia and southward to India. The northern forms evolved into early races with long unbranched stems whereas the southern forms evolved into late-races with lot of branching in the stem.

The chief fiber-flax producing regions are the north temperate regions (70%) of Russia, Poland, the Baltic States, Belgium, Holland and France. Some varieties are grown in warmer regions of India, U.S.A., Canada, Argentina, Uruguay and Russia.

Flax occupies the second or third place in usefulness amongst the fibers cotton and jute ranking first and second. Flax is a much better fiber than cotton and gives rise to a finer fabric.

Linseed is grown mainly for fibers in the temperate zone and for oil in the tropics.

The older Soviet Union produced about 70% of the total world's production of flax.

Belgium and coastal Western Europe produce flax of the highest quality.

The fibers of flax occur within the **pericycle** in the form of bundles, which are about 30 in number. A bundle consists of 10 to 40 fibers. Flax fibers are very fine as well as durable. The fibers have about 65% cellulose, 17% hemicellulose, 2% lignin.

Cultivation

The fiber-yielding variety of flax is taller and grows to a height of 4 feet. They are cultivated in cool and moist (summer) climates of the temperates. The plants are spaced very closely so that the plants become long and slender and remain unbranched. The plants may be harvested before the seeds are matured.

Extraction of Fibers

Retting. The plants are retted in two ways. The plants are usually exposed to dew for two to five weeks during which period the soft tissues decay. This process is called '**dew retting**'. In '**cold water retting**' the plants are submerged in water of a pond, river, etc. for one or two weeks. Retting is quicker in water kept at controlled temperatures. Retting is completed in as short as 2 to 3 days and the quality of the fiber is also better. During retting calcium pectate of the middle lamella dissolves away as a result which the fibers are loosened.

Scutching. The retted plants are dried and the woody core is fragmented, crushed and beaten to separate it from the fibers.

The fibers (group of individual fiber cells) are 1 to 3 ft. in length. They appear yellowish in colour after extraction. The fibers are very fine–tough, flexible, strong, durable, long and lustrous. They are stronger than cotton or wool. The cell wall contains 90 per cent cellulose. The fibers become stronger after absorbing water. They have qualities of strength, durability, gloss, water absorption and drying. The fiber is spun into quality yarn in the spinning mills.

Uses of Flax Fiber

1. The fibers are chiefly used in the manufacture of linen cloth and thread. The threads are very strong and are, therefore, used for tailoring, shoemaking, manufacture of fishing lines and nets etc.
2. The fiber is extensively used in the manufacture of canvas, twine, carpets, cigarette paper, writing paper, firehouses, etc.
3. Fragments of fibers separated during the scutching process are called 'scutching tow' and are used in the manufacture of coarse fabrics and ropes.

HEMP

(Vern. Bhang)

(Cannabis sativa)

Origin and History

The true hemp ***(C. sativa)*** is one of the oldest of the cultivated plants. It is of Asiatic (central and western) origin. The plant is believed to have been under Chinese cultivation more than 4,500 years ago. According to **Schultes** (1970) hemp might have originated in a diffuse manner in a huge area extending from **Caspian** and **Himalayas** to **China** and **Siberia**. The plant was under cultivation in China more than 4,500 years ago. The crop is believed to have spread to West Asia and Egypt during the period 1000 – 2000 B.C. and reached Europe during the Middle Ages (about 1500 B.C.). Later it colonised the Mediterranean region. Hemp fabric has been found in Turkey, (700–800 B.C.). ***Cannabis*** is indigenous to **western and central Asia**. Today it is widely grown in the tropical and temperate regions of the world. Though ***Cannabis*** is a well known **hemp** fiber yielding, crop today it is mainly cultivated illegally as a narcotic yielding plant. It is now chiefly grown in Russia. and Europe (Italy, Poland, etc.). India, Pakistan, Bangla Desh, Nepal, the countries of western Asia, China, Japan, African countries, southern states of U.S.A., Chile and Peru. Italian hemp is considered to be of the best quality. The varieties yielding best quality fiber were introduced in the western world from China.

Hemp is grown widely in temperate zone. Russia accounts for one–third of the total production of hemp in the world. Romania, Hungary, Poland, Czechoslovakia are major producers of hemp in Europe. There are about 10–35 fiber bundles. A bundle consists of 10–40 individual fibers.

There are three kinds of varieties of hemp under cultivation:

(i) varieties grown for the fibrous stems,

(ii) varieties grown for seeds which are edible and yield a drying oil,

(iii) varieties grown in the Indian subcontinent and countries of western Asia for medicinal and narcotic properties.

The plants yield **'marijuana'** or 'hashish', **'ganja'**, and **'bhang'** which have a hypnotic and hallucinating effect. They are derived from resinous secretions of the foliage and the inflorescence.

Botanical Description

The hemp plant belongs to the family **Cannabinaceae**. The plant is an annual herb. It grows to a height of 3 to 16 feet. The slender stem is branched and bears large palmately compound leaves. The plant is dioecious. Male plants yield fibers of better quality. It bears palmate leaves. The plant is dioecious.

The surface of the whole plant is covered with a large number of glandular hairs. The bracts around the female inflorescence has very large number of such glandular hairs. The glandular hairs are rich in resinous narcotic.

The fiber occurs in the pericycle. It is white or dark brown. The fibers are similar to those of flax but are much longer–3 to 15 ft. in length. They are also stronger and more durable than the flax fiber. The flexibility and elasticity of the fiber is, however, less because of its lignification. The cell is composed of about 80 per cent of cellulose. The fiber is unaffected by water.

Cultivation

Hemp grows well in a mild humid and warm climate. It requires an annual rainfall of 30 to 35 inches. Fertile loamy soil with rich amount of humus gives a very good yield of hemp.

It is usually cultivated by sowing the seeds with the help of a drill. The plants for fiber purpose are closely spaced so that they remain unbranched. The plants are cut just after they start flowering to get very strong and flexible fibres.

The fibres are extracted in a way which is exactly similar to that of flax fibre. "Dew retting" and scutching are done before further processing.

Extraction

Narcotic from ***Cannabis*** plants is obtained in three ways.

Bhang. Though the resin content is low, the entire plant is crushed in a bowl like container made of iron with a solid pastle like iron rod. The extract is either consumed along with water or milk or even in the form of ice creams, 'burfi' etc. It is known as **marijuana** in America.

Ganja. The female inflorescence, which is rich in resin content is dried up, crushed in the form of powder and is smoked like tobacco.

Charas or **Hashish.** It is the pure resin secreted by the leaves and the inflorescence of the specially cultivated female plant. The resin content of the secretions is 35% to 45%. Since the resinous secretion has a high concentration of physiologically active tetra hydro-cannabinol (THC) it causes hallucination etc.

Uses of Hemp

1. The fiber obtained from the stem is used in the manufacture of cordage (ropes and twines), sacking, carpets, tarpaulins, thread, binder twine, bags and webbing, etc.
2. The fine quality of the fiber is used in the manufacture of coarse fabrics which looks like a coarse linen cloth.

3. The short fibres *(***tow***)* are used in the manufacture of caulk (**oakum**), and as a packing for pumps, engines, etc.
4. Hemp oil is drying oil obtained from the seeds of the plant.
5. Hemp yields alkaloids which have medicinal and narcotic properties.

SUNN HEMP

(Crotalaria juncea)

Origin and History

It is a very common bast fiber of India second only to jute. It has originated in India from where it has spread to several **tropical countries.** In India, Uttar Pradesh accounts for more than 40% of the total output. The other states producing sunn hemp are Bihar, Madhya Pradesh, Andhra Pradesh and Tamilnadu.

Sunn hemp is widely cultivated for green manure and fiber since it is a fast growing legume. India is a major exporter of sunn hemp fiber to U.S.A, U.K. and Belgium.

Botanical Description

Crotalaria juncea is an annual herb. It belongs to family **Fabaceae**. It is erect and tall attaining a height about 3 to 9 feet. The roots are of tap root system. They bear nodules which fix atmospheric nitrogen. The leaves are alternate, small and sub sessile. The flowers are bright yellow coloured. Inflorescence is a terminal raceme. The fruits are oblong and inflated pods. The seeds are black coloured and are kidney shaped.

There are several varieties of sunn hemp some of which are (i) White, (ii) Green, or Ganjam, and (iii) Dewghuddy. White variety accounts for about 60% of the total production of sunn hemp in our country. The green variety of the fibers have more strength as compared to the white variety. The Dewghuddy variety is creamish in colour and has good length. It is the best variety of the three varieties but is grown on a very limited scale.

Cultivation

Sunn hemp is a hardy plant and can withstand drought conditions. It is grown in hot climates and even on poor soils. The preferred rainfall is upto 75 cm. The land is well prepared before the onset of monsson. Harvesting of the crop for obtaining fibers is done at the pod formation stage, when the plants are about three months old. The plants are either pulled out or are cut at this stage. If the cutting is done at a late stage the fibers become rough but become stronger. The cut plants are left in the field for some time so that their leaves dry up and fall off.

The dried plants are tied into bundles and then thrown into ponds for retting. The retted plants are taken out of water after a week or so and are then spread out in the open for drying. The fibers are then extracted manually before they are dispatched to the factories to be cleaned and used for making a variety of materials.

Location of fiber. Sunn hemp is a **bast fiber** like jute. It is chiefly made of cellulose (80%) and lignin (20%). They are elongated thick–walled fibers in which the lumen is completely obliterated.

Uses of Sunn Hemp

1. The fibers being rough and strong are used as **cordage fibers.** They are used for manufacturing twine, ropes, canvas, sail cloth, sacks, fishing nets, etc.
2. These fibers are also used in making tissue paper, high quality paper, cigarette paper, etc.

RAMIE

(Boehmeria nivea)

Origin and History

Ramie is popularly known as China grass. It is indigenous to China, Japan and the Malaysia. In China it was cultivated in 2200 B.C. China is the chief producer of ramie. It is also grown in Japan, Taiwan, the Philippines, Malaysia, India, Indonesia and Brazil. It is cultivated in the temperate and sub-tropical climate.

Botanical Description

Boehmeria belong to the family **urticaceae.** Ramie is a perennial shrub. It is 6 to 7 feet in height. It is called a **wonder weed**. It has an underground root stock from which arises 2 to 4 aerial stems. The stem bears petiolated heart-shaped leaves which have alternate phyllotaxy. The flowers are unisexual. The plant is monoecious. There are two varieties of ***B.nivea***. The variety **nivea** is called **true ramie**, **white ramie** or **China grass**. It yields fibre of superior grade. The var. ***tenacissima*** is known as **green ramie** or **rhea**. While ***nivea*** variety is cultivated in the temperate zone the ***tenacissima*** variety is grown in the tropical climate of Africa, Mexico and Brazil.

Ecological Factors

Boehmeria is adapted to a variety of climatic conditions, ranging from cool to hot and humid climate. The plant requires an annual rainfall around 100 cm which is well distributed. The sandy loam soil is ideal for cultivating the crop. After each harvesting, the soil requires a high dose of fertiliser to restore the nutrients.

Cultivation

The crop can be raised from seeds but since the plant takes two years to develop fully the crop is generally cultivated by means of cuttings of rhizome and roots. The crop can be maintained in the field upto 6 years. The perennial roots bear new shoots every time the stem is cut.

The young shoots of shorter length yield fine soft fibers whereas long stems yield coarse fibers. Harvesting is done when the flowers are emerging. Ramie is a **bast** fiber. It occurs within the secondary phloem.

Extraction of the Fiber

The fibers have a heavy deposit of gum which makes the extraction process quite difficult. The peripheral portion of the stem i.e. the bark is removed manually or by machines. The fibrous strands are then separated by removing the unwanted pulp etc. The coating of gum is removed by treating the bundles by chemicals, soap or lime steaming. The clean fibers are **"filasse".**

Ramie is the longest fiber. It can attain a length of 55 cm. It is the toughest vegetable fiber and its tensile strength is several times more than any other fiber. When wet it develops much more tensile strength.

Uses of Ramie

1. Ramie is a cordage fiber and is, therefore, not highly desirable in textile industry.
2. In China the fiber is woven for preparing clothing which is popularly called **'grass cloth'** or **"Chinese linen".**
3. Ramie is used for making cordage, threads, twines, nets, sail cloth, sacks, table cloth, water hoses, towels, napkins, cigarette paper, bank notes etc.
4. **Kenaf** (Vern. Patsan). The fibre is obtained from ***Hibiscus cannabinus*** grown in India, Java, Iran, Nigeria, Natal and Egypt.

Surface Fibers

COTTON

(Vern. Kapas)

(Gossypium spp.)

Origin and History

Cotton is the most important fiber in the world. The fiber is obtained from the seed hairs (**lint**) of a few of the several species (about 20) of ***Gossypium***, which are native to both the Old and the New World. Cotton is known to have been under cultivation for weaving of its fibers as early as 4,500 to 5.000 years ago, both in Peru and in the Indus Valley of India (Fig. 12.2). It finds references in the Rig Veda and in the laws of Manu. The Hindus were the first people to use cotton fiber in the form of cloth. *Gossypium* species are believed to have had several centers of origin.

Kinds of Cotton

It is very difficult to classify the numerous wild and cultivated species and varieties of cotton. The majority of species are found in the wild state and only four species are cultivated.

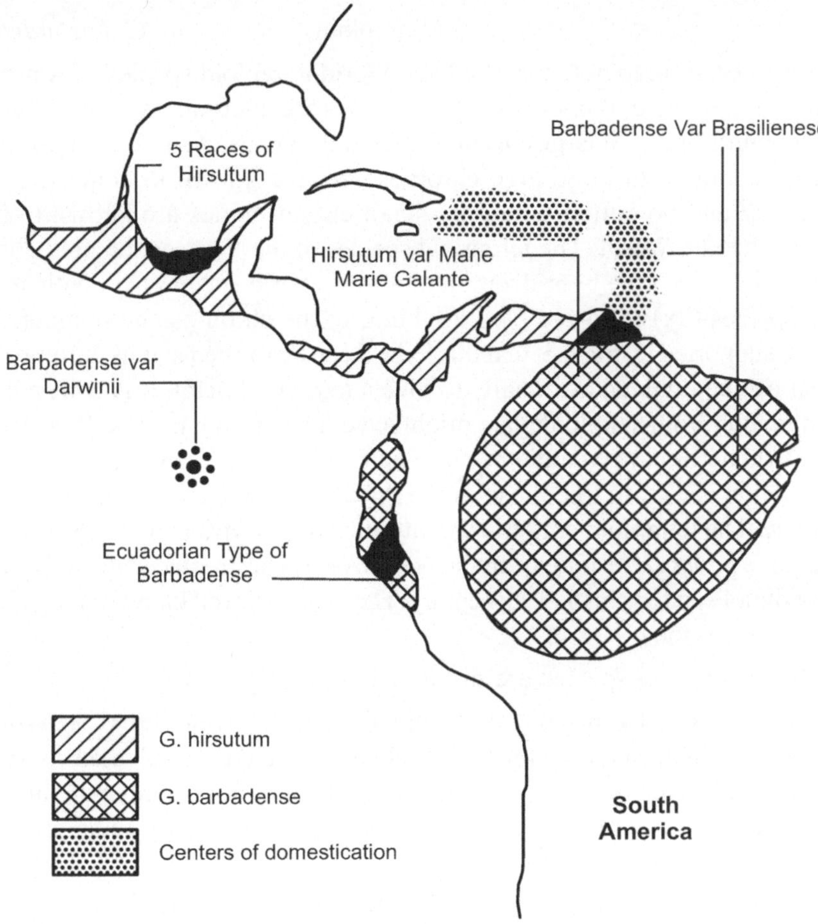

Fig. 12.2. Distribution and centers of origin of the New World tetraploid cottons, ***G. hirsutum*** and ***G. barbadense***.

The cultivated and wild species are either **diploid** or **tetraploid** with haploid chromosomes of 13 and 26 respectively. The tetraploids include only the **New World cottons** whereas all other species are diploid. The diploid species are usually classified into 7 sections depending on their place of origin and the particular genome they possess. The tetraploid species which are only three are kept in a separate section. The cultivated species belong to only two of the 8 sections, *i.e.,* to section VII *(*'**Herbacea**'*)* and VIII (**Hirsuta**) as shown below:

Section VII ('Herbacea'):	Old World Cottons:	*(i)* ***G. arboreum.***
	(Diploid)	*(ii)* ***G. herbaceum.***
Section VIII (Hirsuta'):	New World Cottons:	*(i)* ***G. hirsutum.***
	(Tetraploid)	*(ii)* ***G. barbadense.***

It is interesting to note that the New World tetraploid species of American origin do not grow in the wild state and, therefore, their ancestry is unknown. They contain 13 pairs of large chromosomes and 13 pairs of small chromosomes. Since no cotton with only large chromosomes are known to come from the New World and no cotton with only small chromosomes are known to have grown in the Old World. The tetraploid species of the VIII section are believed to have arisen from the hybridization of a New World species (2x) with an Old World species (2x) followed by a doubling of the chromosome number. This theory is supported by the fact that this hybrid can be formed and its chromosome number can be doubled artificially to give a tetraploid plant. It is believed that the cotton boll from a diploid plant might have floated from the Old World to the New World across the Atlantic because half (larger) the chromosomes of New World cotton are more like those of the African G. *herbaceum* than Asiatic *G. arboreum.* The tetraploid cotton were introduced into Africa in the 18th century by the slave traders returning from the New World. They replaced the indigenous inferior diploid cottons in the 19th century. They are now well known as 'Egyptian' and 'Sudan' cottons.

The *Gossypium* species are discussed below:

Section VII ('Herbacea'). It includes ***G. herbaceum*** and ***G. arboreum,*** which occur wild throughout the Old World tropics and the cultivated cottons of the Old World have been derived from them. The wild species are perennial but the cultivated ones are annuals. The lint hairs produced by their seeds are smaller and are thus of poorer quality.

G. herbaceum is probably of **African origin**. It is extensively cultivated in India, Iran, China, Japan and parts of Africa. The plant is a shrub with rigid stems. It attains a height of $1 - 1\frac{1}{2}$ meters. The bracteoles are rounded or broadly triangular with margin divided into a number of triangular teeth. The cotton bolls are rounded. Their surface is smooth or is very shallowly dented. There are few or no oil glands. The boll splits up into three or four loculi, each containing up to eleven seeds. The cotton is used for low quality fabrics: carpets and blankets. It can be very easily blended with wool.

G. arboreum (tree cotton) is either of **Asiatic** or **African origin**. It is grown in India, Persia and Africa from times immemorial. In the wild state it is a perennial, much branched shrub. It grows to a height of more than two metres. The leaves are five to seven lobed. The bracteoles are triangular, longer than broad, entire or with three or four coarse teeth. They invest the bud, flower and the boll very closely. Bolls are tapering and are profusely pitted with prominent oil glands present in the pits. They open widely when ripe and contain up to seventeen seeds per locule.

Section VIII ('Hirsuta'). This includes the tetraploid species of the New World– ***G. barbadense*** and ***G. hirsutum***. They are American in origin and have spread to every cotton-growing country of the world because of the superior quality of these cottons. They constitute the bulk of the cultivated cottons of the world.

G. barbadense. It is variously called Sea Island cotton, Egyptian cotton, Brazilian cotton, Peruvian cotton. It is believed to be a native of tropical **South America (Brazil)**. Later it spread to the islands of West Indies and was, therefore, named 'Sea Island' cotton. It includes a large group of cottons which produce the finest quality lint. It is cultivated in the West Indies, South America, U.S.A., Egypt and Sudan.

The plant is a tetraploid of the New World. It is a perennial shrub but is treated as an annual under cultivation. The plant may attain a height of nine feet. The flowers are deep yellow and have a red or purple spot at the base. The corolla are not widely expanding and form a narrow cup. The bolls are long, tapering, shiny, dark green, and are profusely pitted with oil glands at the base of the pits. The fruit has three to five locules. The bracteoles are large and have ten to fifteen long acuminate teeth at the tip. They partially enclose the bud flower, or the boll. The seeds are fuzzy only at the ends and testa is covered with long, shining lint hairs. Though ***G. barbadense*** includes the best quality cotton its cultivation is much less than that of ***G. hirsutum***, the American Upland type. The two important types produced by this species are briefly outlined below.

Sea Island Cotton. It is cultivated largely in the West Indies, South America, and in the coastal areas of South Carolina, Georgia and Florida. It produces the finest lint in the world. The lint hairs are creamy white and are silky and lustrous. Staples are of more than 2 in. in length and are stronger and firmer than any other cotton. It is used for the finest textiles and yarns.

Egyptian Cotton. It is grown under irrigation in the Nile Valley of Egypt and Sudan. It was introduced there from Central America. Egyptian cottons are of very mixed origin. The lint hairs are similar to those of Sea Island, but are shorter, measuring 1-3/8 to 1-3/4 in. in length They are brown coloured and are not so fine. This cotton is used for thread, underwear, hosiery, tire fabrics and fine dress goods.

G. hirsutum. It is believed to have originated in **Guatemala** or **Southern Mexico** of Central America. It is the so-called 'American Upland' cotton. It has about 1,200 varieties. It is grown throughout the great cotton belt of the southern United States, in India, East and South Africa and in Sudan under rain cultivation.

The plant is a small shrub with very few vegetative branches. Leaves are occasionally palmatisect. Flowers are white or light yellow and unspotted. Petals are widely expanding. Bracteoles have numerous teeth like projections and closely invest the bolls. The bolls are rounded and often large. They are pale green in colour. The surface is smooth. Oil glands are few and are sunk beneath the boll surface. The bolls are four or five-valved. The seeds are covered with short white **fuzz** and **lint hairs** of staple length of 1/2 in. to 1 in. (the 'short staple American Uplands') or of 1 in. to $1\frac{1}{2}$ in. length(the 'long staple American Uplands').

The lint is strong but not as fine or lustrous and strong as the lint of *G. barbadense.* It produces medium quality yarns and forms the bulk of the cotton trade of the world.

Botanical Description

Gossypium belongs to the family **Malvaceae**. The plants are herbaceous, shrubby or tree like. They are perennial under wild conditions but the cultivated species have developed into annuals. The plants usually grow to a height of 2 to 6 feet.

The plant has a long tap root, varying greatly in length depending on the type of soil and whether the crop is grown under conditions of rain cultivation or irrigation.

The main stem is monopodial, erect and branched. The stem and the vegetative branches bear spirally arranged leaves. The leaves bear two kinds of buds—axillary and extra-axillary. Usually only one of the two buds develop. The axillary buds of the lower nodes produce vegetative or monopodial branches whereas the extra-axillary buds of the upper nodes develop fruiting or sympodial branches. The fruiting branches may also be borne by the vegetative branches. They are truely sympodial since a flower is borne at the tip and the growth is carried on by an axillary-bud of a lower leaf. The fruiting branches are also different from the vegetative branches in the fact that they bear two row of leaves in opposite manner in two rows. The leaves of cotton plant are variable in shape and size. The leaves are stipulate and have long petioles. The lamina is broad, palmately-lobed with three to nine lobes.

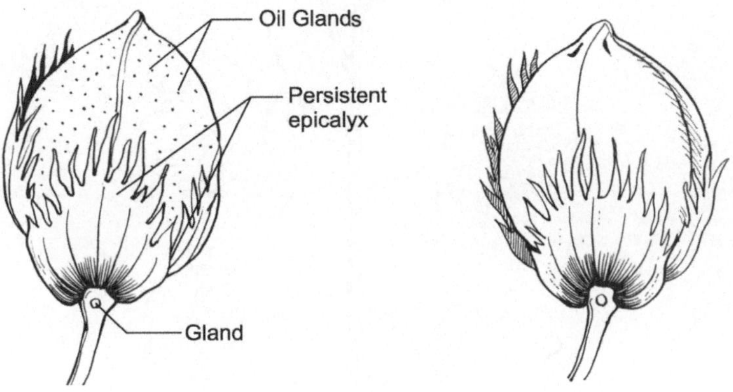

Fig. 12.3. A. Boll of ***G. harbadense;*** B. Boll of ***G. hirsutum***.

The flowers are borne singly from the terminal buds of the sympodia of the fruiting branches. The flowers come to lie opposite the leaves, Normally, cotton is self-pollinated but cross-pollination may also take place.

The fruits of cotton are called 'bolls'. The bolls are variable in shape and size. They are more or less egg-shaped and are pointed at the upper end. The bolls of G. *barbadense* are dark green, profusely pitted with large number of oil glands (Fig. 12.3A). The bolls of G. *hirsutum are* light green, smooth and have few oil glands (Fig. 12.3B).

The cotton fruit is a capsule with three to five locules which are commonly known as 'locks'. When the locks separate the fibrous seeds are exposed (Fig. 12.4). Each locule bears about nine seeds, and a boll may contain twentyfour to fifty seeds. The seeds are ovoid, more or less pointed, dark brown structures. The seed's coat produces two kinds of tubular outgrowths or hairs from the epidermal cells. The longer hairs which are thick and white or creamy are called the '**lint**' and the very short, white or coloured hairs are termed the '**fuzz**' (Fig. 12.5). While the lint or 'staple' is the cotton fibre of commerce, the fuzz is of little importance. While most of the species have fuzz all around the seed, some have fuzz at only one end of the seed, and some do not have them at all. The fuzz hairs are stongly attached to the seed coat. The long tubular cell of the mature lint hair has a very thick wall consisting of superimposed layers of spiral bands of cellulose. The hair is twisted at the points of "pits" which are thin spots in the wall. The number of twists in the fiber

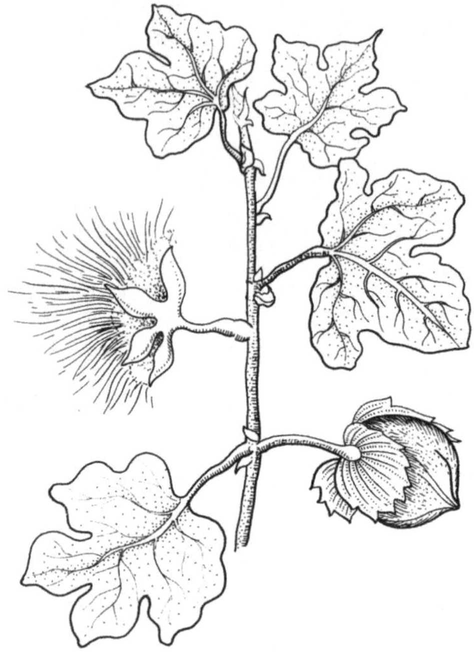

Fig. 12.4. A fruiting branch of cotton showing opened cotton bolls.

varies with different varieities. In the Indian cottons the average number of twist is about 150 per inch, whereas some of the best cottons may have as many as 300 twists per inch. The occurrence of twists in the cotton fibres enable them to be spun in to yarn and is responsible for the coherence of the individual fibres.

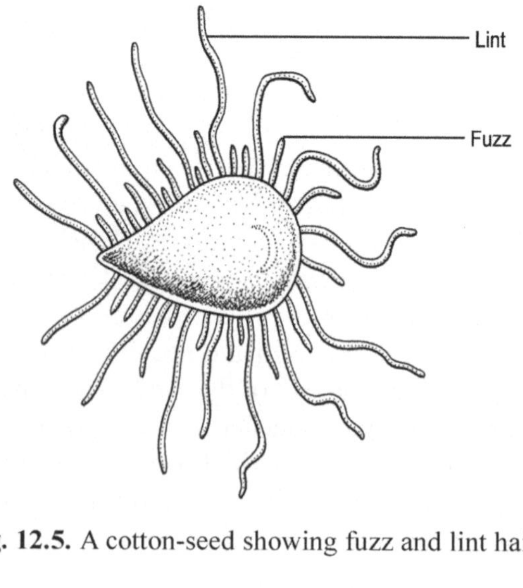

Fig. 12.5. A cotton-seed showing fuzz and lint hairs.

New Varieties of Cotton: SH 131, SH 274, SH 2674, SH 175, 468–10, Varalakshmi, DCH-32.

Quality of cotton fibres

The quality of cotton fibres depends upon several characters. The following characters are important in determining quality:

1. The strength of the fiber, that is, the actual strain that a fibre can withstand.
2. The length of the fiber: The longer fibres are better than the shorter ones. The Sea Island cotton is, therefore, the best cotton grown in the world.
3. Uniformity of the fibres: The fibres of uniform length are easily spun into yarns and there is no undue wastage during manufacture.
4. The number of twists: The fibers with numerous twists make the strongest yarn.
5. Fineness: The fibers with very thick walls are 'coarse' and with thin walls are silky and 'fine'.
6. Colour and lustre: Cottons with rich bright creamy colour are better.
7. 'Counts': The quality of cotton is usually evaluated in terms of counts. A count is the number of hanks required, to weigh a pound, A hank contains 849 yards of thread. Finer threads, therefore, have greater counts. Indian cottons are of about 22 counts whereas the best quality cottons may range from 80 to even 400 counts.
8. 'Neppiness': During the process of ginning some fibers are entangled into knots due to failure of individual fibers to thicken during growth. This is a troublesome and undesirable character.
9. Ginning percentage: This is the amount of lint obtained in relation to the amount of cotton fed into the gin. Obviously the variety of cotton which gives a higher ginning percentage is better.

Ecological Factors

Cotton is a tropical crop and requires adequate but not excessive moisture during early stages of growth, a relatively dry period during flowering and no rainfall during harvesting. It requires a high and uniform temperature during the early stages of growth. There should be plenty of moisture in the soil which can be supplied by rain or irrigation. In the later stages of its growth the temperature should be lower preferably with cool nights. This checks vegetative grewth and encourages fruiting. Dry weather at the time of boll opening and harvest is good. Sandy and moist soils are generally good for its cultivation. The river valleys are, therefore, very good regions for its growth. In India cotton is cultivated in the black cotton soils of the Peninsular India, red soil of southern India and alluvial soils of the Indo-Gangetic plains. In India cutton is grown in 9.13 MH and the yield is 21.04 MT.

Cultivation

Different cultural methods are adopted in different soils and climatic conditions. Deep ploughing is generally not done. The land is usually harrowed 2 or 3 times to facilitate conservation of moisture in the soil. The seeds mixed with mud or cowdung are sown broadcast or in rows by drills. The distance between the lines depends upon the variety grown and the fertility of the soils and usually varies between 12 in. and 36 in. While American cottons are grown pure, the local cottons are grown as a mixed crop. Rotation of crop is beneficial. In northern India (Punjab, U.P. and Bihar) it usually alternates with wheat and jowar, in the black cotton soil tracts of peninsular India jowar precedes cotton. After germination the plants are thinned to 12 to 24 inches apart in the row. During thinning operation, 'roguing' or the removal of off-type plants is carried on.

In black cotton soils when the plants are about 4 inches high a process called interculture is practised. The soil is cultivated with a bullock drawn **daura** or hoe to prevent damage from cracking. Weeding is done by hand in our country. In America weeding is done by tractor-drawn cultivators, flame throwers, and chemical herbicides. The plants start flowering in 7–10 weeks time.

Harvesting is done by hand or by machine. The seed cotton, is usually picked from the open bolls. In America mechanical 'strippers' or 'pickers' are used. A defoliant is applied shortly before harvest to facilitate mechanical stripping.

Preparation of Cotton Thread

This involves a number of operations. They are as follows:

(i) *Ginning.* The lint hairs are removed from the seed and fuzz by means of a saw-tooth gin or a roller gin. The cotton seed, to which the fuzz is still attached but from which most of the lint have been removed is subjected to a further ginning process, and is then used either for seed or is sent to the oil press for oil extraction.

(ii) *Baling.* The lint hairs got from the ginning process are baled under pressure in bales for transporting to the mills.

(iii) *Picking.* In the textile mills a series of beaters and pickers fluff the fiber and frees it of foreign particles and delivers the cotton in a uniform layer.

(iv) *Lapping.* Three layers of cotton fibres are combined into one.

(v) *Carding, combing* and *drawing.* The short fibres are extracted, others are straightened and evenly distributed.

(vi) *Twisting.* The fibers are twisted into threads.

Uses of Cotton

1. Cotton alone or mixed with other fibers are used in the manufacture of various types of textiles.
2. Absorbent cotton consisting of thoroughly cleaned fibres which are almost pure cellulose are basic raw materials for a number of cellulose industries.
3. It is used in the manufacture of rubber-tire fabrics, in plastic reinforcing, carpeting and cordage.
4. Unspun cotton is commonly used for stuffing purposes.
5. Cotton seed oil is one of the important semi-drying oils (25%). It is used for cooking purposes and in soap making.
6. The seed cake is a very useful cattle or stock feed. It can also be used as a nitrogenous fertilizer, and as a dyestuff.
7. The stalks of the cotton plant can be used as a fuel. Thay yield a fiber which can be used in paper making.
8. The linters (fuzz) are used in stuffing pads, pillows, cushions, mattresses, etc. They also yield cellulose, absorbent cotton and low grade yarn. The yarn is used in the manufacture of twine, rope and carpets etc.
9. The **hulls** are a good stock-feed. They are used as fertilizer, and as a source of industrial solvents, explosives and many other things.

COIR

(*Cocos nucifera*)

Origin and Distribution

Coconut palm is believed to have originated in the **Indo–Pacific** region. Today it is worldwide in distribution since the fruits of coconut being light and fibrous have spread over all the coastal regions of the continent by ocean currents. The plants are growing in abundance in the coastal areas of the tropics and the subtropics. Philippines, Indonesia and India stand 1st, 2nd and 3rd in production of dried kernel of the fruit. Other countries growing coconut on a large scale are Mexico, Papua New Guinea, Malayasia, Sri Lanka, Tanzania, and Thailand.

In India coconut palm cultivation is restricted to Kerala, Tamil Nadu, Karnataka, Andhra Pradesh, Orissa, West Bengal, Andaman and Nicobar, and Goa.

Cocos nucifera belongs to the family ***Arecaceae***.

Location of the Fiber

The fiber of coconut popularly known as coir is obtained from the fibrous husk (mesocarp) of the fruits. The fiber strands are about 1 ft in length. Individual fiber cells are approximately 2 ft in length. The fiber is exceedingly light and elastic. The fiber cells are rich in hemicelluloses and lignin. The lignin keeps on increasing with increase in age and, therefore, the fibers become stiffer and stronger.

There are three types of fibers. They are termed **Mat, Curl** and **Bristle** fibers. The best quality coir is obtained from green fruits. In India mat fiber is largely produced in the coastal areas of Karnataka and Kerala. Curl fibers are obtained from unretted husk. It is short and is, therefore, not good enough to be used for making yarn. Bristle fiber is used for making brushes, and brooms.

Harvesting and Processing

The coconut fruits are harvested when they are 10–11 months old. The nuts are split into 2–4 pieces.

Retting. The husk is then allowed to remain in saline water for 8–9 months. The husk is cleaned and is then beaten with wooden stinks to loosen the fibers. The fibers are then allowed to dry up in the open sun. The fibers are manually separated from each other with the help of steel combs. In Kerala coir industry has flourished due to large number of lagoons available for retting the husk. The husk can also be retted by **mechanical** or **chemical methods.** India is the chief exporter of coir and coir products to several countries particularly U.S.A. and European countries.

Uses of Coir

1. Superior quality yarn is used for manufacturing mats, carpets, bags of various sizes and a variety of other articles.

Other Surface Fibres

1. Inferior grade coir is used for manufacturing twines, ropes, etc.
2. Coir is today widely used in the mattresses.
3. Coir is used as a packing material to protect the contents.
4. Coir is used in making writing boards, high strech paper.
5. Coir is rubberized for making cushion seats for automobiles.
6. Coir is used for sound proofing of auditorium, etc.
7. **Kapok.**

Ceiba pentandra is believed to have originated in **tropical America**. Today it is found growing in the forests of India, Sri Lanka, the Philippines, Central West Africa, Malaya, and Java.

Kapok is popularly known as Silk cotton tree. It belongs to the family **Bombacaceae.** The plant requires heavy rainfall. It is a huge tree attaining a height of 50-100 feet. The bark of the tree may or may not be spiny. The tree is generally highly branched. The roots are of buttress type. The leaves are compound. A twig of plant having inflorescence in shown in Fig. 12.6. The fruits are typically ellipsoidal capsule. The fruit bears large number of black coloured seeds. The seeds bear numerous silky floss which is known as kapok. Hairs are unicellular and silky. They arise from the inner walls of the capsule. The fibers are very light and impervious to water.

Fig. 12.6. Kapok Inflorescence.

The floss is produced in the pods of the kapok *(Ceiba pentandra)*. It is a very good silk cotton and is extensively used as a filling fibre. It is used in stuffing pillows, cushions, furniture, mattresses, etc. The fibres are impervious to water and are five times more buoyant than cork and are, therefore, used in the manufacture of life-belts, life buoys, portable pontoons, etc. They are also used for insulating refrigerators and for making soundproof rooms. They have a number of other uses as well. The plant is grown in Java, the Philippines, Sri Lanka, India and Mexico. It grows wild in tropical America.

3. **Red silk cotton** (Vern. Simal). It is called Indian kapok. The fibre is obtained from ***Salmalia malabarica***. It is used for stuffing purposes.
4. **White silk cotton.** It is also native of India. The plant (***Cochlospermum religiosum)*** is cultivated in the tropics.
5. **Madar** (Vern Ak). The fibre is obtained from ***Calotropis gigantia.***

Hard or Leaf Fibres

There are a number of plants which yield hard fibres. They are as follows:

ABACA

(Manila hemp)

(Musa textilis)

Origin and Distribution

Manila hemp is indigenous of Philippines. Today it is grown there as a plantation crop. It is a wet tropical crop requiring heavy rainfall, throughout the year. In India it is grown in Andamans island. Its cultivation is restricted to Indonesia and some tropical countries in America.

Botanical Description

Manila hemp is obtained from wild banana, *Musa textilis* of the family **Musaceae**. The leaf sheaths of the plant are the source for the fiber. Musa textilis is a moderately tall plant with an underground rhizome, a pseudostem made of leaf sheaths and many large sized spreading out leaves. The plant multiplies by means of **'suckers'**, though the plant can be raised from seeds or cuttings of the roots. The plants require a good amount of fertilisers. *Musa textilis* is preferably replanted every 12-15 years though the plant can thrive till the age of 25 years.

Location of the Fibre

Abaca fibers occur within the peripheral part of long leaf sheaths of the plant. The fiber strands are approximately 5 to 12 feet in length. The individual fibers are 2.5 to 12 mm in length. The fibers have been reported to be rich in cellulose (63%), and hemocellulose (20%). Some amount of lignin (5%) is also present in the cell wall. The fibers are the most durable of the structural fibers. It is light and elastic. The fibers are, stripped from the leaf sheaths, cleaned from the pulp and dried.

Uses of Manila Hemp

1. Abaca is highly resistant to saline water and is, therefore, widely used for making marine cordage for the ships.
2. The fibers are used to make twines, ropes and cables.
3. The fibers are used for making fishing nets.
4. The ropes are used in mining.
5. The fibers are also used for making tough paper for tea bags etc.

SISAL

(Agave sisalana)

Origin and Distribution

Sisal is obtained from ***Agave sisalana***. It is indigenous to **Mexico.** It is a tropical plant and is widely cultivated in **Tanzania** which is the largest producer of the crop in the world. Other countries involved in cultivation of species of ***Agave*** are Brazil, Kenya, Angola, Madagascar and Haiti. In India the plant is cultivated in Kerala, Karnataka, Orissa, Maharashtra, Assam and Bengal.

Botanical Description

Agave sisalana belongs to the family **Amaryllidaceae**. The plant has a short and thick stem. It bears a large number of leaves giving the appearance of a whorl. The plant produces approximately 300 leaves during its life of about 10 years. The leaves are very typical in appearance. They are thick, leathery, rigid and fleshy and are bluish green in colour. The leaves are very long with a sharply pointed tip. A mature plant bears a long inflorescence Fig. 12.7. *Agave* is monocarpic i.e. flowers only once. The plant multiplies by means of **bulbils** which develop on floral stalks. The plant also multiplies by means of suckers.

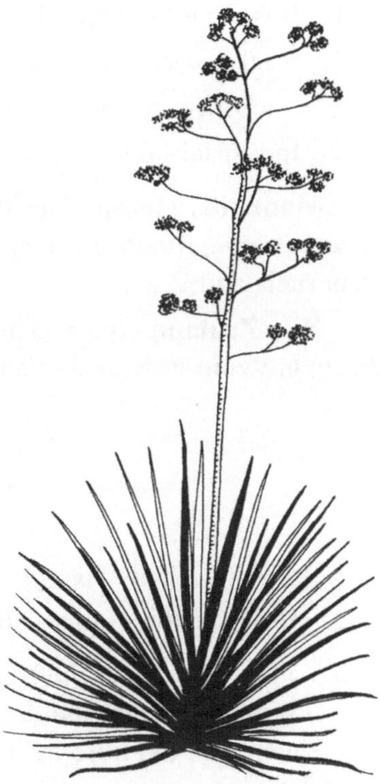

Fig. 12.7. Sisal. Plant with inflorescence

Ecological Factors

The plant has been found to grow well in deserts and extremely arid conditions. It needs high temperature and moderate rainfall of about 100 cm. The plant has been reported to thrive in sandy-loam soils as well as dry black cotton soils.

Harvesting and Processing

When the plant is about 2 years old the leaves of the plant are cut near their bases.

The leaves are tied into bundles of about 30 and are immediately sent to the processing units.

The fibres must be extracted within two days of harvesting. The fibres are extracted manually by beating the cut leaves or by subjecting the cut leaves to high-speed decorticators. The separated fibres are then dried up in the open.

Fiberc The fibers of ***Agave*** are coarse. They are about 3 feet to $4\frac{1}{2}$ feet long. The individual fibers are 2-5 mm long. They are polygonal in shape. The components of the fibers are cellulose (72%) and lignin (14.5%). The fibers are very strong and cannot be decomposed by microorganisms.

Uses of Sisal

1. It is a cordage fiber is mostly used for making twines.
2. Sisal is used for making ropes, bags, matting, paper etc.
3. The fibers are also used for making brushes.
4. Industrial fabrics are also made from the fibers.

Mauritius Hemp. The fibre is obtained from the leaves of green aloe ***(Furcraea giganted)*** growing in Mauritius, Madagascar, South Africa, India, Venezuela and Brazil.

New Zealand Hemp. The fibre is obtained from the leaves of ***Phormium tenax*** growing in New Zealand.

QUESTIONS

1. Discuss the details of different species of cotton.
2. Describe the origin distribution and botany of *Gossypium* species.
3. Name *any five plant* species which yield economically useful fibres. Write a note on the use of these fibres and indicate the morphological nature of the fibre in any *two.*
4. (*a*) Mention the botanical name and economic uses of jute. Where does it grow in India?

   (*b*) Give the botanical name of any fibre-yielding fruit and indicate the use of the fibre.
5. What is fibre? *Name four* fibres of commercial importance in India and write botanical names of the plants which yield them. In each case mention the family to which the species belongs, the plant part from which the fibre is obtained, and the economic uses of the fibre.
6. What is a fibre? Name three fibres of commercial importance that are grown in India. Write the botanical names of the plant which yield them, the family to which each belongs, the plant part from which the fibre is obtained and the economic use of the fibre.

13

Forest Products: Wood and Cork

Forests are known to give a variety of products which are very useful for the mankind. In this chapter only two of the numerous forest products, wood and cork, will be discussed whereas the others are dealt with in other chapters.

WOOD

Importance of wood. The importance of wood to the mankind is next only to the food and clothing. Processed and unprocessed wood finds several uses in our day-to-day life. As a raw material for construction purposes it is highly versatile. It also gives several conversion products. It is cheap, light, stiff, strong, hard and tough material.

Structure of Wood. Botanically the term wood is applied to the secondary xylem produced by the cambium in the stems of gymnosperms and dicotyledonous angiosperms.

The wood is classified into **soft wood** and **hard wood** which are also termed **nonporous** or **porous.** In Gymnosperms the wood usually consists of tracheids and is termed soft wood. The tracheids perform the function of both mechanical support as well as conduction. In Angio-sperms, the wood is termed hard wood and it consists chiefly of two kinds of elements—wood fibres or tracheids which give mechanical strength and vessels, which carry on conduction of water. Besides these elements the living parenchyma *cells* present in the wood perform the function of distribution and storage of carbohydrate food.

The wood of gymnosperms lacks vessels and is, therefore, called non-porous. In angiosperms the wood contains vessels and is, therefore, called porous. The porous wood is of two kinds:

(*a*) **Ring porous** in which the vessels are larger in size and are arranged in a ring. This type of wood is found in mulberry, teak, etc.

(*b*) **Diffuse porous** in which all the vessels are of smaller size. They are scattered within the secondary xylem. Diffuse porous wood is common in the wood of the trees of the tropical climate This type of wood occurs in mango, shisham and sal etc.

The wood has different types of grain, figure and texture. Grain of a wood indicates the arrangement of growth rings, rays, etc. and the wood is either straight-grained, spiral-grained, wavy-grained or curly-grained depending upon the position of the grain. The figure of a wood is the pattern created by different arrangements of grains. The wood is fine-textured or coarse-textured on the basis of the small or large size of the cells.

Wood devoid of moisture is stronger, more durable and rigid as compared to moist wood. Seasoning of wood is done by keeping it in open air, at light temperature or in steam chambers to make it moist-free. Seasoned wood is particularly useful in the construction of furniture, door, show-case, etc.

Uses of Wood

There are a number of uses of wood. They can be broadly classified as follows:

I. Unprocessed timber

1. Lumber and sawed timbers
   - A. Construction, including maintenance and repair
   - E. Crating and shipping supplies
   - C. Manufacture of furniture etc.
2. Fuel wood
3. Posts, mine timbers, poles and pilings
4. Veneers, plywoods and laminated structures
5. Ties and railroad maintenance
6. Cooperage (barrels, tubs, etc.)
7. Miscellaneous items

II. Wood conversion products

1. Pulp and paper
2. Rayon
3. Plastics
4. Modified woods
5. Hydrolysis, fermentation and other hemurgic products
6. Distillation.

A brief account of the above-mentioned uses of wood is given below:

I. Unprocessed Timber

1. Lumber and sawed timbers

Wooden logs cut for use are usually called 'lumber'. Large sized lumber fit for heavy construction or carpentry is called 'timber'. An account of some timber plants is given at the end of this chapter. Lumber is used for a number of purposes.

A. Structural Purposes. Sawed timbers obtained from large-sized lumbers of preferably softwoods are used for the construction of residential and other buildings, sub-buildings, bridges and other types of heavy construction.

B. *Planning-mill* Products. Big lumber is recut into smaller sizes for constructing doors, blinds, window frames and interior finish like baseboards, columns, stairwork, posts, porch work, grills, etc. in the planning mills.

C. Grating. Timber is very widely used for manufacturing crates, boxes, baskets, etc. for vegetables, canned goods, farm products, cosmetics, tea packets, machinery and instruments.

D. Manufactured Products. A number of items are manufactured from wood. They include furniture, railway wagons, vehicles, boats, musical instruments, matches, toys, picture frames, agricultural implements and numerous other items.

2. Fuel

Wood is extensively used for fuel particularly in rural areas. It is a very good fuel since in dry condition it is 99 per cent combustible.

Besides wood, fossilized remains of plants in the form of **coal** *is* also a very widely used fuel. Coal gas, benzol, naptha, ammonia, coal tar (source of dyes, antiseptics, etc.) and coke are obtained from coal by destructive distillation. **Lignite** is also used for fuel purposes.

Accumulated deposits of vegetable matter in the bogs and swamps are slowly decomposed and carbonized and are termed **peat**. It is also used for fuel in some countries.

Petroleum, a product of ancient plants and animals is an important source of energy. Refined petrol is used in motor cars, scooters, and in industries. High quality gasolin obtained from fractional distillation of petroleum is used as aviation fuel. The other products are kerosene, jelly and paraffin.

3. Posts, mine timbers, poles and pilings

Wood is widely used for fencing posts around farms, along roads, railways, etc.

Wood, mostly in the form of unsawed timbers, is used as **props** are supporting overhead burden in tunnels as **lagging** between the props and the walls to prevent their collapse and as **ties**, (foundations for rails).

Non-conductive and non-corrosive cylindrical, durable sapwood **poles** are used for telephone, telegraph, electricity and power-transmission lines.

Pilings, preferably softwoods, are meant for bearing heavy top loads. They are used in building constructions, bridges, restles and docks.

4. Veneers, plywoods and laminated structures

Veneers are thin sheets of wood cut from softened (boiled) logs *by* **rotary, slicing** or **sawing** process. Plywoods are three or more (always an odd number) sheets of veneer glued together at right angles to one another.

The cheapest quality of veneers are called **container veneers**. They are used for boxes, barrels, crates, baskets, etc. The **commercial** veneers are used for plywood, furniture, radio cabinet etc. **Face veneers** are used for finest work. The plywood is used for panelling walls, flooring, partitions, doors, cabinets, shelves, furniture and a variety of other things.

5. Ties and railroad maintenance

Uniformly sawed wooden slippers or ties treated with preservatives are extensively used below the rail tracks throughout the world. Though metallic ties are also being used these days, the wooden ties continue to be preferred because of their cheapness, elasticity, resistance to shock, strength, durability and their easy replacement.

Such woods are also used in the manufacture of freight cars, trucks, and other similar vehicles.

6. Cooperage

Wooden staves are bound together for manufacturing a variety of containers. There are two kinds of **cooperage—slack cooperage** and **tight cooperage**. Slack cooperage is meant for holding and transporting solid substances like flour, fish, meat, vegetables, fruits, tobacco, crockery, sugar, cement, etc. The tight cooperage includes barrels and kegs which are meant for transporting liquids such as beer, wine, oil and other liquids. Paraffin is applied to the inner surface to prevent any leakage. Heavy cooperage includes huge tanks and vats.

7. Miscellaneous items

Excelsior. Excelsior or **wood fibre** are thin strands of wood used for packing and shipping glassware and other breakable articles and for stuffing mattresses and upholstery because of their lightness and resiliency. They are also odourless and free from dust and dirt. Fine quality excelsior called **wood wool** is used in the manufacture of rugs and matting.

Sawdust and **shavings** are used as fuel and as packing material.

Wood flour produced by grounding wood wastes, sawdust and shavings is used in the manufacture of linoleum, plastics, explosives, veneer bonds and many other products.

Shingles and **shakes** are thin sections of wood arranged in an over lapping manner on the roofs and sides of buildings to protect them from the weather.

II. Wood Conversion Products

PAPER

Origin and History

Paper has a very interesting history. The ancient **Egyptians** in 2400 B.C. used the pith of a plant ***Cyperus papyrus*** for preparing a paper like vegetative parchment called '**papyrus**' from which the word 'paper' is derived. The Romans are also known to have used the Egyptian papyrus for writing purposes. The first makers of paper were, however, the Chinese. The paper industry spread to India, Persia, and Arabia from China. Later it spread to Spain and other European countries. In 1690 a paper mill was established in U.S.A. The Chinese made paper from pulp prepared by boiling old rags, silk waste, hemp, rice leaves and bark of mulberry trees. The invention of a machine by French worker **Louis Robert** in 1799 for manufacturing paper and the discovery by **Frederic Keller** in 1845 that ordinary wood could be converted into a paper pulp led to the establishment of the modern paper industry.

Raw Materials

The chief raw materials of paper are natural fibrous materials like wood fibres, cotton and linen. Cellulose in the pure form gives the best quality paper.

Wood pulp composed of long and strong cellulose fibers is the most wanted raw materials for the paper industry. **Spruces** ***(Picea spp.)***, **pines** ***(Pinus spp.)***, **aspens** ***(Populus spp.)***, and the **balsam fir** (***Abies balsamea)*** are important wood pulp-yielding plants. Some hardwoods like **beech** ***(Fagus grandifolia)***, **sugar maple** ***(Acer saccharum)*** and **birch** ***(Betula lutea)*** also furnish wood pulp.

Cotton and linen rags, which were the only source of paper before the wood pulp became established as the chief raw material, continue to be used for producing fine papers. Cotton has a very high percentage of cellulose (91%) and linen (flax fibre) contains 82% pectocellulose.

Other raw materials of minor importance are textile fibres of jute and hemp, wastes of the textile industry, the esparto grass (***Stipa tenacissima)***, paper mulberry, papyrus *(Cyperus papyrus)*, bamboo ***(Bambusa spp.)***, baobab ***(Adansonia digitata) Dephne cannabina*** and ***Eulaliopsis binata*** (Baib-grass) and a number of other plants and fibrous materials. Low-grade paper, strawboard, cardboard and pasteboard etc. are produced from the fibrous stems of wheat, rye, barley, rice, oats and other grasses.

Manufacture of Wood Pulp

The wood pulp industry is very large. The conversion of wood into pulp is brought about by mechanical or chemical processes.

The mechanical process is employed in the manufacture of poor-grade papers which are not very good in strength, durability and printing. This paper contains resins, lignin, etc. because of which it turns yellow in colour. The newsprint is manufactured by this process.

The chemical processes remove lignin and other undesirable things. There are three kinds of chemical processes *(i)* **soda process**, *(ii)* **sulphite process** and **sulphate process**. Soda process is the oldest one and is practised for the reduction of hardwoods and pine by alkaline cooking. The sulphite process involves cooking of wooden chips with steam in a solution of acid calcium sulphite. It is employed in processing softwoods of low resin content into a very strong pulp. 50 per cent of the total pulp is formed in this way. The pulp gives high quality paper. The sulphite process is now widely used for producing pulp from coniferous woods with a high resin content. The pulp is used in manufacturing a strong brown wrapping (kraft) and fibreboard containers.

The Manufacture of Paper from Pulp

The pulp is first bleached and washed. The pulp is then subjected to beating in a machine called **beater** or '**Hollander**' to separate the fibers. Sizing to make the paper water and ink resistant and dye, if any, is added during this process to colour the paper. The pulp is then transferred to the paper making machine where it passes through a number of stages before yielding paper. The machine consists of very long wire screen revolving around a series of rollers.

Kinds of Paper. Though paper is primarily used for writing and printing purposes, a number of paper products are also used in our daily life. The different kinds of paper and paper products are as follows:

(i) **Fine papers**. The finest papers are made from rags or sulphite pulp. Mixtures of rags and chemical pulp are used for medium quality papers. The cheaper quality is made from wood pulp only.

(ii) **Newsprints**. As stated earlier it is made from wood pulp prepared by the mechanical process.

(iii) **Wrapping paper**. It is obtained from the coniferous woods and some textile fibres.

(iv) **Blotting paper**. Soda pulp alone or in a mixture or cotton rags yield blotting paper.

(*v*) **Paper board**. Waste materials like boxes, old newspapers mixed with mechanical or chemical pulp yield paperboard.

(*vi*) **Vegetables parchment**. Waterproof and grease proof parchment is prepared by dipping paper in sulphuric acid. It is used for packing purposes.

(*vii*) **Carbon paper**. Cotton or linen paper is coated with a coloured wax. It is used for copying purposes.

(*viii*) **Facial tissue paper**. Sulphur pulp is used for manufacturing this paper.

(*ix*) **Cigarette paper**. It is made from flax.

(*x*) **Roofing and building papers**. The paper is treated with wax, tar, oils, resins or other substances.

(*xi*) **Papreg**. The paper is mixed with plastics. It is moulded into a very strong but light weight material.

RAYON

Cellulose derived from chemical pulp and cotton linters (short hairs removed after ginning in the cotton mills) is extensively used in the mills to yield another textile fiber, or yarn which is called rayon or '**artificial silk**'. The cellulose is dissolved and the resulting liquid is then drawn into long strands or threads which are hardened to give the rayon fibres. Chemically rayon is, therefore, cellulose or an ester of cellulose.

There are four kinds of rayon produced from four kinds of chemical processes each differing from the other in the solvent used for dissolving the cellulose. They are *(i)* **Viscose rayon** *(ii)* **Cellulose acetate rayon**, *(iii)* **Cuprammonium rayon**, and *(iv)* **Nitrocellulose rayon**. The first, third and fourth are pure cellulose while the second one is an easter of cellulose.

Rayon yarn is extensively used as a textile fibre and is second to only cotton in importance in the manufacture of cloth. It can be more easily dyed than the natural fibers. It is used in the manufacture of various household and industrial textiles.

Plastics and Other Products

Cellulose nitrate formed by the action of concentrated nitric acid on cellulose gives a variety of useful products. They include **celluloid**, **plastics**, **guncotton**, **pyroxylin**, **collodion**, **varnish**, and **artificial fibers**. Wood, wood wastes, and wood flour are used in the manufacture of plastics. The **lignin** is believed to be responsible for the physical properties of plastics and other similar substances. The first plastic celluloid, was made from cellulose nitrate and natural camphor.

Today plastic products occupy a very important place in our daily life. A large variety of plastic articles are manufactured these days. Celluloid is used in the manufacture of billiard balls, toys, handles, etc. and is dyed to give a number of colourful objects.

Modified Woods

Wood is treated with resins and is then heated to setting temperatures and is finally compressed in presses to give modified woods. The modified woods are dark coloured, heavy, hard and durable but are also very brittle. They are used for panelling, airplane propellers, pulley and gear wheels, insulators, etc.

Hydrolysis

The cellulose of the wood wastes, sawdust, etc. is hydrolysed to convert it into simpler carbohydrates (sugars), which can be fermented to give rise to ethyl alcohol, acetone, butanol, lactic acid, etc. The wood can also be directly hydrolyzed into methanol, acetic acid and furfural.

Distillation

Softwood is distilled to produce turpentine and rosin. Hardwood is distilled to produce charcoal, acetic acid and wood alcohol (methanol). Charcoal is used as a fuel in many countries.

Cork

Cork is obtained from the bark of trees, particularly from cork-oak ***(Quercus suber)***. The tree grows in the **Mediterranean region**, which is its native place. The cork-oak growing countries are Spain, Portugal, southern France, Italy, Algeria, Tunisia, Morocco and Corsica.

The tree is usually 20 to 60 ft. in height and has a diameter of 4 ft. It grows well on rocky slopes of the mountains and survives in the extremely dry conditions because of the cork.

The cork cells of the stem are produced as a result of the meristematic activity of the cork-cambium ring. Their cell walls have a deposit of waxy **suberin**. They are impervious to water and enclose air in a water-tight condition.

Cork is removed in the form of a continuous strip of a thickness of 1 inch. The first stripping is done when the plant is about 20 years old, and the subsequent strippings are done at intervals of 8 to 11 years. The tree usually lives for 100 to 500 years.

The cork pieces are first of all dried in the open and then steamed or boiled to remove the sap and tannins. This also increases their volume and elasticity.

The outer rough portion is removed by hand and the curved sheets are pressed to make them flat. The sheets are then trimmed, graded and baled.

Cork has a variety of uses. It is used as bottle corks. It is widely used in the manufacture of insulating board, linoleum, hats, handles for golf clubs, floats, life jackets, mats and countless other articles. It is used in sound-proofing rooms, etc.

TIMBER YIEDING PLANTS

The gymnospermic **softwoods** are commercially more important since they have some fine properties which make them highly useful. Some of their properties are as follows:

(*a*) They are light.
(*b*) They are easy to handle and work.
(*c*) Their wood is interchageable because they have similar wood properties.
(*d*) They grow usually in pure stands and are, therefore, easily cut.

The angiospermic **hardwoods** have the following **disadvantages**.

(*a*) They are heavy.
(*b*) They are difficult to handle, transport and to work.
(*c*) There is great diversity in the qualities of wood and, therefore, this wood is not interchangeable.
(*d*) They hardly grow in pure stands. Only few plants grow in a highly mixed type of forest.

Softwood is usually sold as **sawnwood** for construction; **roundwood** for packaging purposes, for fencing; **veneerwood** for plywood and **pulpwood** for paper and board industries. The hardwood has also similar uses. High quality hardwood like that of teak is mainly used for decorative veneer work and furniture.

The important softwoods are ***Pinus spp***., ***Picea spp***., ***Larix spp***., ***Pseudotsuga spp***., ***Cryptomeria japonica*** and ***Araucaria spp***., The hardwoods include ***Betula spp***., ***Eucalyptus spp***. ***Fagus spp***., ***Populus spp***., and ***Tectona grandis***.

TEAK
(Vern. Sagwan)
(Tectona grandis)

Distribution. The teak tree is native to central and-southern India and southeast Asian islands. It grows well south of 25° N latitude in India, Burma, Malaysia, Indonesia, etc. In our country it is cultivated throughout Bengal, Assam, Sikkim and north-west India and Madhya Pradesh (Nagpur region). It belongs to the family **Verbenaceae**.

Morphology. It is a large deciduous tree. The branches are four sided. The leaves are born in opposite manner. They are long, petiolate, elliptical, rough and glabrous. The plant bears small white flowers.

Ecological Factors. Teak grows well up to a height of 3,000 ft. above which it has poor growth. It thrives well in mean annual temperature of 72° to 81°F. It requires a mean annual rainfall of 50" to 120". It grows on a great variety of soils. It grows on light and sandy soils, as well as on those which are heavy. Perfect drainage and a dry subsoil are absolutely essential for teak plantations.

Quality and Uses of Teak Wood

Teak is the most durable of the commercial timbers of the tropics. In Burma, the tree is usually seasoned by a girdling process in which a deep circular cut is made through the brown-grey bark, the sapwood and the heartwood. The girdled tree dies after a few days. It is allowed to stand for one or two or even more years for seasoning.

An outstanding quality of a seasoned teak timber is that it does not split, crack, shrink, warp or alter its shape. The wood contains an aromatic oil which is believed to be responsible for the great durability of the teak wood. It works easily and takes a fine polish. It does not corrode in contact with metal.

The teak wood has a variety of valuable **uses** which are as follows:

1. It is used for shipbuilding, boats, masts, etc.
2. It is widely used in construction of houses, bridges, piles, pit props in coal mines, railway wagons and carriages, railway sleepers, ploughs, and for cooperage work, etc.
3. Teak wood is prized for high quality furniture, cabinet work, ornamental veneers, bowls, toys, bodies of harmonium and piano and for numerous other things.
4. Teak plants have other uses as well. The leaves are used as plates for thatching etc. The leaves yield a red dye. The wood yields a pain-relieving substance. Flowers and the young fruits are believed to be diuretic. An oil is obtained from the teak wood which is used as a varnish and as a substitute for linseed oil.

SAL

(Shorea robusta)

Distribution. The sal tree is second only to teak in commercial importance in the Indian sub-continent. In India there are two distinct sal belts which are separated by the Gangetic plain. The northern sub-Himalayan belt extends from Kangra valley in the Punjab to Assam. The second or central Indian belt extends from the Panchmarhi hills in the north-west through Rewah, Chota Nagpur, Midnapore to the Godavari river in the south. It belongs to the family **Dipterocarpaceae**.

Morphology. It is a large tree. The tree is branched. It bears leaves which are glabrous, long and broad-ovate. The plant bears yellowish flowers in racemes in large, branched panicles.

Ecological Factors. The plant grows up to a height of 3,000 ft. It requires a mean annual rainfall between 40 and 100 inches and a temperature range between 55° and 88° F. The plant grows well on loose and porous soils.

Quality and Uses of Sal Wood

The sapwood is not durable but the heartwood is very strong, heavy and tough. It has a cross-grained structure.

Sal timber takes a long time for seasoning and tends to split and warp with the change of dry and wet seasons. It is a very durable wood but not as much as the teak wood.

The various uses of the sal timber are as follows:

1. The timber is widely used for pit-props, bridge construction, railway sleepers, piles, boat building, masts, oars, poles, dyeing vats, carriages, cart wheels, ploughs, rice pounders, etc.
2. Sal is also used for manufacturing inferior quality furniture, etc.
3. It yields charcoal which is used by blacksmiths.
4. The bark of the tree is used as a tanning material.
5. The stem excludes an aromatic oleoresinous gum '**sal dammar**' which is used for caulking boats, in shoe polishes, carbon papers, typewriter ribbons, paints, varnishes, etc. It is also burnt as an incense. Sal dammer yields 'chua' oil on distillation. This oil is used for preparing 'attars' and incenses. It has some medicinal properties as well.
6. 'Sal butter' is obtained from the seeds. It is used as an illuminant, and as a substitute for cocoa butter in the manufacture of chocolates.

SHISHAM

(Dalbergia sissoo)

Distribution. It is native to the sub-Himalayan tracts and the outer Himalayan valleys from Punjab to Assam. It also grows in Baluchistan, central India and Gujarat. It belongs to the family *Leguminosae.*

Morphology. It is a large deciduous tree attaining a height of 60 ft. and more. It has a trunk of a diameter of 6–12 ft. It bears compound leaves. Each leaf has 3–5 leaflets. Yellowish white flowers are borne in short axillary panicles.

Ecological Factors. The plant grows to a height of 3,000 ft. and sometimes 5,000 ft. It requires a very moist climate. It usually grows on sand or gravel along the banks of rivers.

Quality and Uses of Shisham Wood

Shisham wood is well known for strength and elasticity. It is widely used for the following purposes:

1. It is very much used in building of boats, carts, carriages, agricultural implements, etc.
2. It is used in the construction of buildings, doors and window-frames.
3. It is a very good furniture-wood.
4. It is used for interior decoration, cabinet work and various carpentary works.
5. It is also used in the manufacture of various items like toys, articles for decoration purposes, combs, 'chappals', sports items, drums, etc.
6. The heartwood yield an oil which cures skin and foot diseases of cattle.
7. The twigs and leaves are used as fodder for the cattle.

SEMAL

(Salmalia malabarica syn. ***Bombax malabaricum)***

Distribution. It is native to **India** and **Burma**. It is usually cultivated. It grows up to an altitude of 6,000 ft. It is also called **red silk-cotton** tree. It belongs to the family **Bombacaceae**.

Morphology. The plant is a very large tree. It attains a height of 150 ft. and a girth of 40 ft. The branches are horizontal and in whorls of 5 to 7. The young branches are covered with conical prickles. The bark is smooth and silvery-grey. The leaves have 5 or 7 leaflets. Large scarlet or white flowers are borne in the axils of leaves. The fruit is an ovoid capsule which breaks into 5 segments. Numerous seeds with fine silky fibres are released after the dehiscence of the fruit.

Quality and Uses of Semal Wood

The wood is greyish white. It is coarse-grained and porous. The heartwood is not distinct. In dry condition the wood is not durable.

It has several **uses** which are as follows:

1. It is used for planking, packing-cases, toys scabbards, inferior quality pencils, cheap furniture, brush handles, box veneers, etc.
2. It is a very good wood for match boxes and match splints.

3. The trunk of the tree is usually hollowed out for making drums, water pipes, fishing-floats and rafts for floating and transporting heavier timbers.
4. The timber is used for piles which are immersed in water.
5. The fine 'pieces of the semal wood called 'wood wool' is used as a packing material.
6. The silky fibres of the seeds are used for stuffing purposes.
7. The seed yields an oil which is used as an illuminant and for making soaps, etc. The cake and the seeds are used as a cattle-feed.
8. The stem exudes a gum (mocharas) which has medicinal properties.

DEODAR

(Cedrus deodara)

Distribution. The deodar is restricted in distribution. It is native to the mountains of Afghanistan and North Baluchistan, and to the northwest Himalayas. The plant is capable of growing to an altitude of 12,000 ft. It belongs to the order **Coniferales**.

Morphology. The plant is a large evergreen tree. It is a tall tree and attains a height of 250 ft. It is very much branched. The tips of the smaller branches are drooping. The bark is dark coloured. Leaves are 1 – 11/2 inch long, right and sharp. Cones are erect.

Quality and Uses of Cedrus Wood

The wood is yellowish-brown. It is oily and aromatic. The sap-wood decays during the seasoning of the logs. The heartwood is very durable and is almost imperishable in the climate of Kashmir and Punjab. The wood is light and keeps well in dry as well as wet conditions. Deodar like teak, shisham and sal is an important timber yielding plant of our country and is put to various uses.

1. The timber is widely used in the construction of barracks and other buildings, bridges, canals, railway sleepers, railway carriages, telegraph poles, etc.
2. It is also used for constructing storages vats for beer, packing cases, boxes, toys, furniture, carts, musical instruments, etc.
3. Destructive distillation of the wood yields 'deodar tar oil' which is antiseptic and is used for preserving skins.

PINE

(Pinus spp.)

Pine timber (a soft wood) is an oustanding wood and accounts for almost one-half of the total lumber supply of the world. The wood is obtained from several species of ***Pinus***, out of which eight species are important timber-yielding plants. Broadly speaking the pines are classified into **soft**, and **hard** types. The soft pine wood is straight-grained, soft, almost free of resin and is easy to work. The hard pine wood is resinous, heavy, hard, strong and durable. The hard pine wood is resinous, heavy, hard, strong and durable.

Some of the important pines are described below:

Pinus excelsa (Vern. **Kail**)

It is also known as **blue pine**. It is a large evergreen tree. It has clusters of five needles per dwarf shoot. It is the best of our pines. It grows in the temperate Himalayas from Bhutan westwards at an altitude of 6,000 to 12,500 ft.

It is second only to deodar in durability. The wood is soft and easy to work. It is pinkish red and is moderately hard and highly resinous.

'Kail' is used for house buildings, shingles, water-channels, water troughs, wooden spades, packing cases, match boxes, match sticks, body of violin, etc. It yields an excellent charcoal for iron-smelting. It yields turpentine, resin and tar. The bark is used for roofs of huts in the forests.

P. longifolia (Vern. **Chil, Chir**)

It is a large tree with symmetrical branches. The leaves are in clusters of three. It grows in outer Himalayas and Siwaliks at altitudes of 1,500–7,500 ft. It is cultivated in the plains of north-west India.

Chir is widely used for house building, oars, furniture, tea chests, boat building, general carpentry, sports items, match boxes, etc. The chir is believed to produce more turpentine and resin than the other coniferous trees of the north-west Himalayas. The crude turpentine and resin obtained by making incisions in the stem or by stripping off the bark are called '**Biroja** or '**ganda firoza**', in north – west India, '**dhup**' *in.* U.P., '**berja**' *or* '**biroja**' in Garhwal, and '**khalija**" near Shimla. '**Biroja**' has medicinal properties. The bark is used for tanning purposes. The bark also yields a dye which is used for dyeing tassar silk into a yellowish shade, and wool into a rich orange shade.

Pinus strobus *(***northern white pine of America***)*

It grows in north-eastern United States and parts of Canada. In England it is called Weymouth pine. The plant is a large tree and attains a height of 100 to 200 ft and a girth of 18 ft. The leaves are borne in clusters of five.

The wood is white or slightly yellowish-white, very soft, light and easy to work. It is durable but has little strength.

In northern America the timber is widely used for doors and windows, furniture, cabinet work, boxes, matches, etc.

Other white pines of America are **western white pine** *(P. monticola)*, the **sugar pine** (***P. Lambertiana***) and the **lodgopole pine** *(P. contorta)*.

Pinus australis *(***southern yellow pine of America****)* It grows in the southern states of U.S.A.

The wood is the best of the softwoods in heaviness, hardness, strength, stiffness and toughness. It is also very durable.

The hard pine is widely used for beams, heavy constructions, bridges, ships, cars, railroad ties, boxes, etc.

QUESTIONS

1. Describe the morphology of teak plant. What are the uses of teak wood?
2. What qualities make teak wood an excellent timber – yielding plant?
3. Discuss the botany of *Cedrus deodara* and *Shorea robusta* and the ecological factors favouring their growth.

14

Industrial Plants
Tannins and Dye Materials

TANNINS

Tannins are complex organic compounds which are astringent and acidic. They are glucosidal in nature and consist of carbon, hydrogen and oxygen like the carbohydrates. They are water soluble. Their exact role in plants is not understood.

They are, however, of great economic importance. They convert animal hides into leather by uniting with their proteins. The leather thus formed is stable, resistant to air, moisture, temperature and bacterial action. Tannins also react with iron salts to form dark-blue or greenish-black compounds, which are the basis of ink. Tannins find use in medicines because of their astringent properties. They are also used in oil drilling to reduce the viscosity of the drill.

The Tanning Industry

Tanning of hides was done about a thousand years before Christ. The tanning of hides is done in a slow and gradual manner so that tannin reaches every part of it. The hide is first dipped in used tannin liquors, the acidity of which causes swelling and thus facilitates the entry of increasing concentrations of tannin. The entry of tannin to the inner part is, however, not possible if the hide is dipped in a strong solution of tannin because the latter causes the leather to become impervious on the other surface. The leather after tanning, is treated with oil or grease, dye, and some finishing materials like gum, resin, albumen, etc.

Sources of Tanning Materials

Tanning is found in almost all the plants but only some species contain substantial amounts. Tannins can be obtained from the following sources:

(*i*) **Bark.** Hemlock *(Tsuga canadensis),* oak *(Quercus spp.),* mangroves particularly red mangrove *(Rhizophora mangle),* wattle *(Acacia spp.),* larch *(Larix decidua)* avaram bark *(Cassia auriculata),* and babul *(Acacia nilotica)*

(*ii*) **Woods.** Chestnut *(Castanea dentata)* and Quebracho *(Schinopsis spp.)*

(*iii*) **Leaves.** Sumac *(Rhus spp.),* Sicilian sumac *(Rhus coriaria)* and Gambier *(Uncaria gambir).*

(*iv*) **Fruits.** Myrobalan nuts *(Terminalia spp.),* Divi-Divi *(Caesalpinia coviaria),* tara *(Caesalpinia spinosa),* algarobilla *(Caesalpinia brevifolia),* and Turkish oak *(Quercus macrolepis).*

(*v*) **Roots.** Tanners dock *(Rumex hymenosepalus),* and palmetto *(Sabal palmetto).*

The Manufacture of Ink

Tannin is the source of another important item of commerical importance—ink. The ink has been in use since 2600 B.C. in **China**. Egyptians are also known to have used ink for writing on papyrus in 2500 B.C.

Of the several types of modern inks, carbon and tannin ink are most important.

Carbon inks. Chinese or **India ink** and **printing ink** are examples of carbon inks. They remain on the surface of paper like paints. China ink is manufactured from soot, carbon black, or lampblack prepared by burning a vegetable oil or pine wood and glue, gum arabic etc. Carbon for the printing ink is prepared from petroleum, natural gas, etc. and is mixed with a drying oil (linseed oil), rosin, some chemical drier and soap.

Tannin inks. Tannin combines with iron salts to give a blue-black colour. Tannin derived from the insect galls formed on the twigs of Aleppo oak *(Quercus infectoria)* is the chief source of tannin inks.

Coloured inks. Aniline dyes or natural dyestuffs are mixed with water, alum and a gum to give coloured inks.

DYES AND PIGMENTS

Vegetable dyes have been in use since times immemorial. Egyptian tombs had colourings of saffron, safflower, indigo and madder The primitive people of Britain decorated their skins with a blue dye. The use of vegetable dyes has, however, declined in the last one hundred years after the discovery by ***Perkin*** (1856) of an indigo-type dye from a 'useless' coal-tar component. The synthetic or aniline dyes obtained from coal-tar products are brighter, more permanent, cheaper, easier to use and afford a wide range of colours. An account of the vegetable dyes is, therefore, of only historical interest. Only a few natural dyes have been able to compete with the synthetic ones.

Dyes are chiefly used in textile industry for giving fast colours to the fabrics, and for colouring paints, varnishes, leather ink, paper, wood, furs, food, cosmetics, medicines and toothpastes.

The dyes are treated with mordants, salts of metals like aluminium, iron, chromium, etc. for making them insoluble. The mordant forms a fine layer of metallic oxide on the fabric. The oxide along with the dye form an insoluble compound which helps the dye to remain fast on the cloth.

Sources of Natural Dyes

Natural dyes are obtained from various parts of a number of plants. The sources are as follows:

WOODS

Haematoxylon campechianum* (Logwood)**: The dye is known as **haematoxylon**, It is obtained from the heartwood of the leguminous tree, ***H. campechianum. The dye is about 10 per cent in content. It is violet, grey, blue or black in shade. The dye is famous in histological studies because of its capability to stain **nuclei**. It is also used in dyeing silk, wool and cotton.

Pterocarpus. The plant belongs to **Leguminosae** family. ***P. santalin*** yields a red dye, **santalin**. The red dye is obtained from the wood and root of the plant. The dye is used to colour idols etc. Natives of Ododobo, Africa used to paint their bodies with the dye. The dye is dissolved in alcohol to give a salmon pink colour to the fabrics.

The dye is also obtained from ***P. tinctorius***. It was used to colour the newly born children by the natives of Pungo Andongo, Africa. The native women also used the dye for colouring their feet to give the appearance of shoes or slippers,

Acacia. The plant belongs to the **Leguminosae**. family. **A. catechu** yields a dye called '**Katha**'. The dye is obtained from the **heartwood** of the plant when it is boiled with water. It is largely used in printing and in colouring pulp and paper. If is also used for dyeing cloth. It is often used with diazo salts to give bright shades. The ship sails are also dyed with Katha to protect them from weather and sea water.

LEAVES

Indigofera. The plant belongs to the **Leguminosae** family. It yields the dye, **indigo** ("neel"). It is called the "**king of dyestuffs**". The dye has been known to the people of Asia for over 4000 years. The word indigo is derived from a Latin word **indicum**. The dye is known to be very much in use in Persia in 600 A.D. Trade in indigo was carried on by Arabs who carried indigo to the Mediterranean region. The dye became very popular in Europe. The dye reached China from Persia. India was a leading producer of indigo during the reign of East India Company and later British Government. During the freedom movement for the liberation of our country from British rule Mahatma Gandhi championed the cause of indigo cultivators of the Champaran district of Bihar against their exploitation by Britishers. ***Indigofera tinctoria*** is the principal source of the dye. 'The plant is a native of **southern Asia**.' *I. arreca* is another important source of the dye.

The leaves of the plant contain a soluble colourless glucoside, called **indican**. During fermentation of the leaves indican is hydrolysed to glucose and indoxyl. **Indigotin** is obtained as a result of oxidation of indoxyl. If the dye is of high quality it is deep violet-blue and porous. The natural dye has now been completely replaced by the synthetic one.

The dye is used in dyeing and printing cotton fabrics, rayon and wool. It is used in washing of cotton clothes. It is also used in the manufacture of pigments for printing inks, lacqueres and paints.

Isatis. A blue dye called ***woad*** is obtained from ***I. tinctoria***. The dye was very popular in Europe before it was replaced by indigo. The dye was called universal dye during the Middle Ages. The cloth used to be dyed blue with this dye and yellow with a dye obtained from ***Reseda luteola***. Such a dyed cloth was called '**Saxon green**'. The plant is a native of **Tibet** and **Afghanistan**.

The leaves of the plant are crushed into a pulp which is then made into balls. The balls are dried and fermented to yield the colouring principle, the indican.

Lawsonia. It belongs to the family **Lythraceae**. A red coloured dye is obtained from the leaves of ***Lawsonia alba*** ('Mehndi). The plant has been known to Indians from the remote past. The dye is favourite with Indian women who colour their palm, feet, fingers with it on festive occasions. The leaves are crushed in the form of a paste for this purpose. The hair is also dyed with it.

ROOTS AND TUBERS

Rubia. The plant yields a red dye. The two species of ***Rubia*** which are important are ***R. tinctorium*** and ***R. cordifolia.***

R. tinctorium (Madder) was known to ancient Indians, Persians and Egyptians. The clothes of the Egyptian mummies were dyed with the dye. Libyan women dyed their dresses with this Alexander got uniform of his soldiers dyed with it.

The roots of ***R. cordifolia*** growing in the Himalayan region yield the dye used in our country. It is used for dying fabrics.

Morinda. The plant belongs to **Rubiaceae**. A yellow dye is obtained from the barks of the roots of ***M. citrifolia*** and ***M. angustifolia.*** The dye is obtained in an optimum amount when the plants are three or four years old. The dye protects paper, cloth, etc. from white ants.

Curcuma. The turmeric plant belongs to **Zingiberaceae**. An orange red dye is obtained from the tuber of ***C. longa.*** The colouring matter is known as **curcumin**. The dye is very important in the tropical countries. It is widely used in our country for colouring food particularly pulses, vegetables, etc. It is used in religious functions of the Hindus and is believed to be very auspicious.

FLOWERS

Carthamus. Safflower ***(Carthamus)*** plant belongs to the **Asteraceac** family. The plant finds reference in Sanskrit works as '**Kusum**' or '**Kusumbha**'. A scarlet red dye is obtained from the florets of ***C. tinctorius.*** The Egyptians dyed the grave cloth for the mummies with this dye. The dye is used in our country to dye cloth for getting cherry red colour. The dye is also used for imparting colour to biscuits, waffers, cakes, etc.

Crocus. Saffron plant belongs to the **Iridaceeas** family. A yellow dye is obtained from the orange coloured stigmas and the upper part of the style of ***Crocus sativus.*** Saffron was well known to the ancient Greeks and Romans. It also finds reference in Sanskrit literature. The plant is grown only in Kashmir valley of our country.

Saffron dye is used for colouring rice and flavouring vegetable dishes. The dye is also used on ceremonial occasions. It is applied on the forehead since it is considered to be auspicious.

Other Plants. ***Butea monosperma*** and ***Nyctanthes arbortristis.***

FRUITS

Terminalia. The plant belongs to **Combretaceae** family. Yellow or black dyes are obtained from the fruits of ***T. chebula.*** In combination with alum the dye gives a permanent yellow colour. Treatment of the dye with iron salts gives black colour.

Other Plant. ***Rhamnus infectorius.***

SEEDS

Bixa. A bright yellow or flesh coloured dye is obtained by macerating the pulp (aril) of the seeds of ***B. orellana.*** It is used for dyeing silk and for colouring butter, cheese, ghee, etc.

GUM—RESIN

Garcinia. It belongs to **Guttiferae** family. The dye, '**gamboge**', is obtained from the bark.

The dye is used in the water colours and gold coloured spirit varnishes for metallic surfaces. The dye is used for dyeing silk cloth.

A similar dye, Japanese dyestuff (Fukuji) is obtained from the bark of *G. spicata.*

LICHENS

Lichens like ***Parmellia omphalodes*** etc. yield dyes (orchil, etc.) which are used for dying clothes. Litmus is obtained from ***Roccella tinctoria.*** It is an indicator of the acidity and is used in chemical studies.

QUESTIONS

1. Write an account of the dyes obtained from the stems of plant.
2. From which part of the plant saffron is obtained what are its uses.
3. Give botanical names, family and botany of any 5 dye yielding plants.

15

Industrial Plants (Contd.)

Rubber and other Latex Products

RUBBER

History. Rubber is indigenous to **South America** (Fig. 15.2). Rubber has a very interesting history. The rubber industry is, however, of a recent development of not more than a hundred years old. Amazonian Indians of the tropical South America are known to have used the legendry 'caoutchouc' (rubber) since prehistoric times for making balls, torches, and water tight containers. The Mexican Aztecs played ball games with balls made of rubber. Europeans noticed this in the early part of the sixteenth century. Spanish settlers in 1600 used latex for smearing shoes, clothes, and hats, etc. for making them waterproof. **La Condamine** (1734) reported *'heve'* or *'jeve'to* be the native name for rubber in the Andean region. Later the rubber tree was given the generic name of **Hevea**. In England **Joseph Priestley** in 1770 discovered its ability to **rub off** pencil marks on paper because of which 'caoutchouc' was now called 'rubber'.

Macintosh in 1823 discovered that rubber was soluble in naptha. This initiated the waterproofing of cloth in places where fresh latex was not available. Sixteen years later in 1839 Goodyear invented vulcanization which completely revolutionized the rubber industry. The earlier trouble with rubber was that it became soft with heat and brittle with cold. During vulcanization rubber combined chemically with sulphur at high temperature (150°C) to become tough and resistant to wear and tear. This most significant discovery and the development of pneumatic tier has established rubber as one of the foremost items of commercial importance.

In Brazil willed trees of ***Hevea brasiliensis*** growing in the **Amazonian rain forests** were cut down to collect rubber. When its great value was realized by the **Brazilian government** it banned the export of seeds. An Englishman, **Farris**, in 1873, however, smuggled out some seeds to the Royal Botanic Gardens at Kew, in London. The seedlings were later shipped to Royal Botanic Garden at Calcutta but failed to survive there and were lost. Two years later, in 1875,

Sir Henry Wickham, under orders from the British government collected about 70,000 seeds along the banks of the Amazon and shipped them in baskets labelled 'delicate specimens for Queen Victoria's gardens at Kew'. 1,900 seedlings were later shipped to Sri Lanka which laid the foundation of the vast rubber plantations of **Sri Lanka, Malaya** and **Indonesia**. Rubber plantations on a large scale have also been established in Liberia, West Africa. It has not, however, been possible to establish plantations in Brazil, the home of rubber, because of certain indigenous diseases and other factors. Malaysia, Thailand, Indonesia, Sri Lanka and India together produce 90 per cent of the world's natural rubber. In India—Kerala, Tamil Nadu and Karnataka are rubber-growing states.

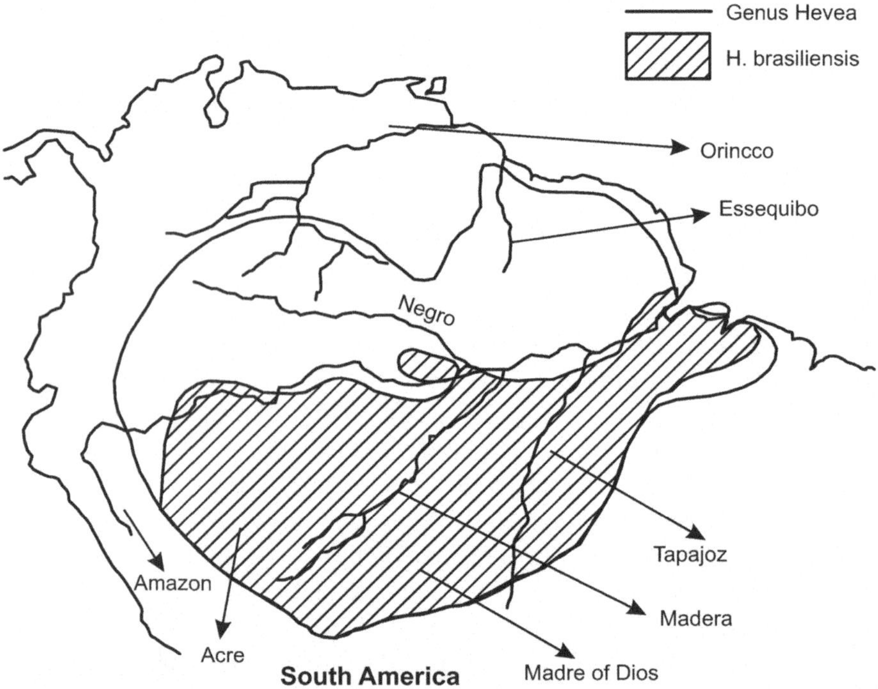

Fig. 15.1. Diagram showing distribution of ***Hevea*** in general and the species ***brasiliensis*** in particular.

Formation of latex in plants. Chemically rubber is an amorphous hydrocarbon belonging to the group of polyterpenes to which the essential oils also belong. It is a constituent of a white or yellow, slightly viscous fluid or **latex** produced by various erect or climbing woody plants of the tropics or subtropics belonging to the **Moraceae**, **Euphorbiaceae**, **Apocyanaceae**, **Papaveraceae**, **Araceae**, **Caricaceae**, **Sapotaceae**, **Asclepiadaceae**, etc.

Latex is produced in special cells (latex tubes) and latex vessels as a byproduct of general metabolic activity. A latex tube is a single cell which grows and branches in a coenocytic manner like a fungal hypha. It becomes multinucleate and grows to enormous size. On the other hand, a latex vessel, like a xylem vessel, is formed by the breakdown of the transverse walls of special cells produced by the meristematic cells of a plant. The tubes and vessels, however, perform a similar function and remain living cells. The exact function of latex in plants is not clearly understood. Latex exuding from the cut surface of a stem, leaf or root seal off a wound. It may also act as a fluid reservoir, in transport of material and for protection and nutrition.

The composition of plant latex varies considerably from species to species and only the latex of some dicotyledons yield rubber. The rubber occurs as minute microscopic particles floating in the watery latex. The particles develop a protective layer of nitrogenous substance to prevent their coagulation. Besides the particles of rubber, latex may contain various acids, mineral salts, sugar, starch, proteins, enzymes, alkaloids, resins, gums and oil etc.

An account of the most important source of rubber, the ***Hevea*** or **Para-rubber tree**, ***Hevea brasiliensis*** is given below.

HEVEA RUBBER

Hevea brasiliensis belongs to the family **Euphorbiaceae**. It is native to tropical **South America** (Amazon Valley) where it grows wild in the hot damp forest. It also grows wild in eastern Peru, Bolivia, Colombia, Brazil, etc. It has been introduced in Liberia, Sri Lanka, Malaysia and Indonesia. The last three countries named here constitute 97 per cent of the total world production of rubber.

Fig. 15.2. Leaves and fruits of para-rubber tree.

The plant grows well within a temperature range from 75 to 90°F and a rainfall of 80–120 inches. The para-rubber tree grow to a height of 60 to 150 ft. The tree trunk grows to a thickness of six feet. The tree bears compound leaves of three leaflets. The flowers are small and are borne in inflorescences. The fruits are three lobed with each lobe bearing one seed (Fig. 15.2). The seed contains about 50 per cent of a dark red drying oil.

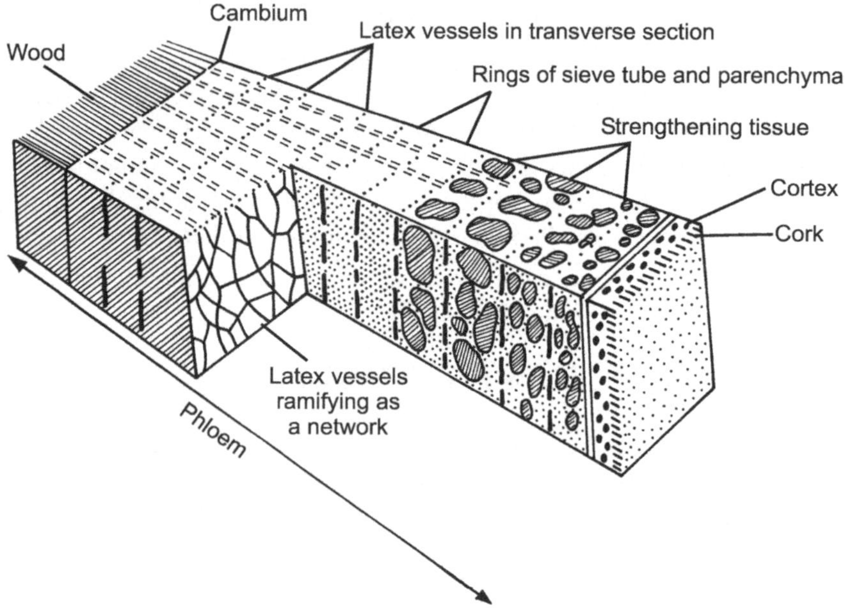

Fig. 15.3. Diagram of outer tissues of para-rubber tree *(H. brasiliensis)* showing the arrangement of rows of latex vessels in the phloem.

Tapping of Rubber. Latex of para-rubber tree contains about 30 per cent dry rubber. The latex is obtained from the numerous latex vessels of the bark which are arranged in concentric rings alternating with rings of phloem (Fig. 15.3). The inner bark contains more vessels than the outer bark. The bark is cut in such a manner that the delicate growth layer of cambium is not damaged. Since the latex vessels run spirally to the right at an angle of 30° of the vertical plane tapping cut is made from the upper left to the lower right at a 30° inclination to obtain maximum yield. Tapping is begun in six-year old trees. First a vertical cut is made in the bark in the lower portion of the tree. Several short downward spirally running incisions which extend either one-third or one-half of the circumference of the tree are made in the bark with a special knife. Sometimes a 'V' cut is made. The latex which flows for several hours in the forenoon is collected in a cup (Fig. 15.4). The flow of latex stops by midday and, therefore, cut surfaces have to be renewed each morning for several months.

The collected latex is sent to the factory to be coagulated by the additions of acetic acid or formic acid to form a white, spongy mass of crude rubber. This is then pressed between rollers and dried to produce sheet of crepe rubber which is smoked in 'smoke houses' at 45°C. In the smoke houses the vapours from burning wood darken the rubber which is now called 'smoke sheet'. The rubber is usually exported in this form and is processed further in the importing countries. Sometimes the latex is centrifuged into a concentrated form and an anticoagulant (ammonium hydroxide) is added to it for export purposes.

Fig. 15.4. A photograph of rubber plantation.

The Amazonians, however, still employ the primitive method of rubber production. The latex around a paddle is allowed to coagulate over a smoke fire. More and more latex is poured over the rotating mass of coagulated latex around the paddle. Layer after layer of latex get deposited to form a huge ball of rubber which may weigh upto 200 lbs. Smoked ball rubber, though produced in a crude manner, is of very good quality.

In the recent past considerable attention is being paid for the improvement of rubber plantations. Bud from a high-yielding clone is inserted on a seedling having vigorous growth to give rise to a high-yielding plant. The plant is then made disease-resistant by inserting another bud to the top from a disease-resistant clone. This **double budding technique** results in a tree which has a vigorous rootstock, high-yielding trunk and disease-resistant top.

In the recent past chemists have been able to produce synthetic rubber. Catalytic polymerization of **butadiene** and **styrene** in soapy water gives a latex which can be coagulated with acid and salt, filtered, and dried to yield synthetic rubber. The synthetic rubber, is now greatly improved. Most of the automobile tyres are now made of a mixture of natural and synthetic rubber.

OTHER SOURCES OF RUBBER

Panama or Castilla Rubber

It is obtained from ***Castilla elastica*** which is native to **Mexico** and **Central America**. Upto 1850 it was of greater importance than the hevea rubber and was the chief source of rubber. The plant belongs to the **Moraceae** family.

Ceara or Manicoba Rubber

It is obtained from several species of ***Manihot*** which belongs to the **Euphorbiaceae**. The important rubber-yielding species are ***M. Glaziovii,*** the Ceara manicoba of northeastern Brazil, ***M. helptaphylla, M. dichotoma,*** etc.. The plants are grown in Sri Lanka, India and many other tropical countries.

Assam Rubber/India Rubber

Assam rubber or **India rubber** is obtained from ***Ficus elastica*** of the **Moraceae**. It is native to northern India and Malaysia. The rubber is of poor quality.

Guayule Rubber

It is obtained from guayule ***(Parthenium argentatum)*** of the **Asteralceae**. It is native to the United States.

Landolphia Rubber

The plants are woody vines of Africa. They belong to *Landolphia* of the **Apocyanaceae**. The important species are ***L. kirkii, L. heudelot's*** and ***L. owariensis***.

Palay Rubber

The rubber is obtained from ***Cryptostegia grandiflora*** and ***C. madagascariensis*** of the ***Apocyanaceae***. They are native to **Madagascar** on the East African Coast. They are also grown in Haiti.

Dandelion Rubber

It is obtained from the Russian dandelion ***(Taraxacum koksaghza).*** The latex is extracted from the tuberous roots of the plant.

Uses of Rubber

1. Three-fourths of the crude rubber is used for manufacturing tyres and tubes of automobiles, aeroplanes, bicycles, etc.

2. Rubber is extensively used in the manufacture of boots, shoes, hose tubes, mats, belting, raincoats and other waterproof materials, toys, insulated wire, etc.
3. Crude rubber is vulcanized with 30 per cent sulphur to produce hard rubber, which is used in telephones, radio parts, surgical appliances, etc.

OTHER LATEX PRODUCTS

Castilla Rubber

(Castilla elastica)

Castilla rubber is indigenous to **Mexico** and **Central America**.

Botanical Description

Castilla elastica belongs to the family **Moraceae**. It grows in rainforests of tropical America. It is a tall tree. It requires less rainfall as compared to ***Hevea brasiliensis***. The pattern of branching is unique in ***Castilla***. Long permanent branches upto 20 ft. long are borne on the main stem. These branches are replaced by another crop of permanent branches after a few years. These permanent branches bear temporary branches which are replaced by new permanent branches.

Flowers are unisexual and are borne on the temporary branches. The plant bears an aggregate type of fruit from the female flowers.

The latex produced by ***Castilla*** is very difficult to extract and is much less than that of ***para rubber***. ***Castilla*** is tapped 4 to 5 times in a year.

GUTTA-PERCHA

Some members of the family **Sapotaceae** yield a greyish-white latex which is converted into a non-elastic rubber called guttapercha. The latex is chiefly obtained from ***Palaquium gutta***. The plant is native to Malaya and grows in Malayasia, Indonesia, Philippines, and other tropical countries.

The latex is obtained either by cutting the trees or by removing barks and making cuts. The latex is collected on banana leaves or in coconut shells. It coagulates immediately.

Gutta-percha is widely used in the construction of submarine cables because of its properties of resistance to salt water, and pliability. It is a very poor conductor of electricity and, therefore, it is used for insulation purposes. It is also used in the manufacture of pipes, golf balls, telephone receivers, adhesives, toys, supports, etc. It protects wounds and is used in dentistry.

BALATA

Balata like gutta-percha is a non-elastic rubber. It is obtained from the latex of ***Manilkara bidentata*** earlier called ***(Mimusops balata)***. This large tree is indigenous to **South America** and **Trinidad**, where it grows in a wild state. The uses of balata are similar to those of gutta-percha. It is specially useful in the manufacture of machine beltings.

CHICLE

Chicle is another example of a balata like latex-product. Chicle is the name of the **gum** which is obtained from the latex of ***Achras zapota*** of the **Sapotaceae**. The genus is native to Yucatan peninsula of **Mexico** and **Guatemala**. It also grows in **Central America**, the West Indies, and in the northern part of South America. It is also a wild large tree. Zigzag cuts are made throughout the length of the trunk and the latex is collected at the bottom. The collected latex is boiled for about two hours in large couldrons after which they are removed and made into firm balls, blocks, etc. This raw chicle is purified, dried and powdered. The powder forms a very sticky mass on heating. Finally, the gum is cleaned, filtered, sterilized and mixed with different flavouring materials to produce the **chewing gum**. Due to a very large demand of chewing gum inferior latex from other members of Sapotaceae is also used in its manufacture. It is very much likely that synthetic plastics will lake the place of 'natural' chewing gum in course of time.

JELUTONG

Jelutong is a gum obtained from the latex of a Malayan tree ***Dyera costulata***. It is very much used as a substitute of chicle. For sometimes it was also used as a source of rubber.

Gums are the familiar exudates of the plants of the dry regions. They are commonly used as adhesives. Besides, they are used in printing, paints, candies, medicines, and for finishing textiles and sizing of papers, etc.

QUESTIONS

1. Give the origin, distribution and morphological features of *Hevea brasiliensis*.
2. How is rubber manufactured?
3. Name four rubber yielding plants and write a line about each one of them.

16

Gums and Resins

GUMS

Gums are produced when cellulose breaks down in the process of **gummosis**. They are rich in sugars and are very much similar to pectins. They are soluble in water.

There are a number of gums out of which gum arabic, gum tragacanth, and karaya gum are very important.

Gum Arabic. It is obtained from wild species of ***Acacia*** particularly from ***A. senegla (A. arabica).*** The genus is native of Northern Africa. The plants grow wild in the dry, sandy, thorn scrub regions of North Africa, Arabia, and India. They are also cultivated in Sudan. The plant is a small thorny tree.

The trees are trapped after the rainy season by making incisions in the bark. The hardened exudates are collected at intervals of 10 days.

The gum is translucent, tasteless and odourless. It dissolves completely in water and yields a highly viscous substance with a high degree of adhesiveness. It is also used in the preparation of polishes, inks, sizes, confections, medicines, etc.

Gum Tragacanth. It exudes naturally or after incisions are made into the bark of ***Astragalus gummifer*** and some other species. It belongs to the family **Leguminosae**. The plant is native of **western Asia** and **southeastern Europe**. Iran and Turkey are the chief producers of this gum.

Gum tragacanth is typically opaque. It is formed from the pith and medullary ray cells of the lower trunk or roots.

The gum is used in calico printing, as an adhesive, and in medicines.

Karaya Gum. It is also known as India gum since it is obtained from a central Indian plant, ***Sterculia ureus*** of the **Sterculiaceae**. Other names of karaya gum are kadaya, katira, kuteera, katilo, kullo, Sterculia gum and Indian tragacanth, The plant is a large tree.

Large incisions are made into the heartwood, from which the gum oozes.

It is a cheap gum. It is used in the textile, cosmetic, cigar, taste and food industries.

Ghatti Gum. It is derived from the stem of ***Anogeissns latifolia*** of the family **Combretaceae**. The plant is native of India and Sri Lanka. The plant is a large tree. The gum is used as a substitute of gum arabic.

Locust Bean Gum. It is variously called carob gum, St. John's bread, gum hevo, gum gatto, lupogum, rubigum, tragon, tragosol, etc. The gum is the exudates of the fruit of ***Ceratonia siliqua*** of **Leguminosae**. The carob tree is widely grown in the Mediterranean region. The gum is a hemicellulose.

The gum is used in paper making, and as stabilizer in food products.

Guar. It is obtained from the guar plants, ***Cyamopsis tetragonoloba*** of **Leguminosae** It is grown in India.

Other Gums Some of the other sources of gums are ***Chukrassia*** *(***Meliaceae***)*, ***Butea spp****. (***Leguminosae***)* and ***Cochlospermum*** *(***Cochlospermaceae***)* of India; ***Feronia*** *(***Rutaceae***)* of India and Southeast Asia; ***Caesalpinia*** *(***Leguminosae***)* of Argentina; ***Prosopis spp.*** (Mesquite gum) of Mexico and South America; ***Moringa*** *(***Moringaceae***)* of the Asian tropics; ***Cycas circinalis*** (Cycas gum) of Asia and a number of other plants.

RESINS

Resins have been used by the ancient civilizations in the remote past. The Egyptians used resins for coating mummy cases. The Greeks and the Trojans used turpentines. The early Chinese and Japanese artistes used lacquer in the paintings.

The manner of resin formation is not clearly known. It is believed to be oxidation products of essential oils. Some resins are supposed to be formed from reduction and polymerization of carbohydrates (starch, etc.). They may be a secondary product of plant metabolism.

Resins occur as secretions in special ducts and canals in different parts of the plants. Sometimes they ooze out through the bark in a natural manner. Commercially, they are tapped by making injuries in the stems or roots of the plants. They are also obtained from fossil materials.

They are insoluble in water but are soluble in ether, alcohol and other solvents. They are almost transparent. They are combustible. Chemically they are very different from the carbohydrate of the gums. They consist of resin, acid, anhydride and ester mixtures. They may exude in the pure state or in combination with essential oils, gums, etc.

The commercially important resins are obtained from the families **Pinaceae**, **Leguminosae** and **Dipterocarpaceae** and **Burseraceae**.

The utility of resins to plants is not clear but they are believed to reduce the rate of transpiration and to prevent the decay of the plants because of their antiseptic properties. Resins are, however, of great importance in industry because of their several properties. As they do not dissolve in water and harden gradually, they are used as **varnishes**. They are mixed with oils or solvents to impart a brilliant, transparent, waterproof layer on the painted surface. Spirit varnishes preferably with hard resins are used as paper size, furniture and leather finishes, and shellac substitutes. The resins dissolve in alkalies to form soap. Resins are also used in making oil cloth, linoleum, floor coatings medicines, fireworks, plastics, incense, perfumes and in many other items.

Synthetic resin like **nitrocellulose**, a derivative of cellulose, is now widely used as spirit varnishes. The synthetic resins synthesized from phenol and formaldehyde, glycerin, urea, vinyl materials, and esters of various other organic compounds are used in paints, varnishes, safety glass, waterproofing of cloth, etc. The natural resins are now finding less use in face of competition from synthetic resins.

It is difficult to have a proper classification of resins because of the use of a variety of terms in the trades, *viz,* varnish resins, hard resins, spirit varnishes, balsams, gum resins, damar resins, soft resins, gums, etc, The resins can, however, be classified in the following three large groups:

Hard resins – Usually solid

Oleoresins – Usually liquid

Gum resins – Mixture of gums and resins.

HARD RESINS

They contain very little essential oils and are, therefore, best sources of varnishes. Moreover, they dissolve in alcohol. They are usually solid, tasteless and odourless. They are combustible and burn with a smoky flame. Copals and damars, the two most important commercial resins fall under this category.

Hard resins besides being very good sources of varnishes are also used in paints, plastics, inks, fireworks, sizing adhesives, etc.

Copals

They are groups of very hard resins. They are obtained from **fossils** and **semifossils**.

West African Copals. They comprise **Congo copal**, **Sierra copal** and **Acera** and **Benin copals**. The Congo copal and Sierra copal are obtained from several species of ***Copaifera*** of **Leguminosae**. Acera and Beni copals are derived from ***Daniella ogea***.

East African Copal. They comprise **Zanzibar copal**, **Madagascar copal** and **Mozambique copal**. They are all derived from ***Trachylobium verrucosum*** of **Leguminosae**.

South American Copals. They are derived from ***Hymenaea courbaril*** of **Leguminosae**.

Kauri Copal. It is obtained from the **Kauri pine** *(**Agathis australis**)* of New Zealand. The plant belongs to' **Pinaceae**.

Manila Copal. It is derived from ***Agathis alba*** of **Pinaceae**.

Damars. They are obtained from members of **Dipterocarpaceae** growing chiefly in Malaysia and Indonesia. The damars are as follows:

Batavian Damar is obtained from ***Shorea weisneri*** of Java and Sumatra.

Mata Kuching Damar is obtained from ***Hopea micrantha***. **Penak Damar** is obtained from ***Balanocapus Heimii.***

Temak Damar is obtained from ***Shorea hypochra.***

The **Indian damars** are **sal damar** *(**Shorea robusta**)*, **white damar** *(**Vateria indica**)* and **black damar** *(**Canarium strictum**)*.

Amber

It is one of the oldest resins in use. It is hard and brittle. It is obtained from the fossils of ***Pinus succinifera,*** found along the shores of the Baltic Sea. Amber is yellow, brown or black coloured.

It is used for making dolls and ornamental items. It is also used for the mouthpieces of pipes and holders for cigars and cigrettes. It also yields varnish and an essential oil. It is used to increase the elasticity of rayon fibres.

Lacquer

It is obtained from the lacquer tree *(**Rhus verniciflua**)* of China and Japan. It is a very good varnish which is used for ornamental purposes. **Burmese lacquer** is obtained from ***Melanorrhoea usitata.***

Shellac

It is not produced by plants, but from an insect ***Laccifer*** *(**Tachardia**)* ***lacca*** which feeds on ***Butea monosperma, Schleichera oleosa, Zizyphus xylopyrus, Ficus religiosa, Acacia nilotica. Cajanus cajan*** and ***Zizyphus jujuba***. These plants are cultivated in India on a large scale for obtaining shellac (97% of the total world output). Burma, Thailand and Indo-Chinese countries account for the rest.

Shellac is very useful in industries. It can be easily moulded into various shapes and is widely used in the manufacture of phonograph records, insulating materials for electrical goods, etc. It is a source of spirit varnish used for interior works since it is not water-resistant. It is also used in the manufacture of sealing wax, drawing inks, water colours for stiffening felt hats and for sizing papers and for decorative purposes.

Other hard resins are acaroid resins ***(Xanthorrhoea spp.)***, sandarac ***(Tetraclinis articulata)***, mastic ***(Pistacia lentiscus)*** Dragon's blood ***(Daemonorops spp.)***, kinos ***(Pterocarpus spp.)***.

OLEORESINS

Oleoresins are liquid in nature due to the presence of essential oils. They have characteristic odour and flavour. Some of the important oleoresins are turpentines, balsams and elemis.

Turpentines

They are obtained from coniferous trees. The crude turpentine obtained by tapping the trees is distilled to yield an essential oil, oil of turpentine or spirits of turpentine, and rosine. The major resin canals are interconnected and run vertically in the sapwood, while smaller ones connect them to the bark.

The turpentine industry is also called 'naval stores' industry because of the widescale use of turpentine in the sailing vessels from early days. Turpentine-yielding plants are ***Pinus palustris*** or ***P. australis*** (90% of total yield), ***P. caribaea***, ***P. pinaster*** and some other species of ***Pinus***.

The spirits of turpentine or the oil of turpentine is known as 'turpentine'. It is widely used as a solvent and thinner in paints and varnishes, in medicine and as a solvent in rubber and guttapercha. It is also used in the printing cloth and in the manufacture of many chemicals.

Rosin or colophoney is a solid. It has many important industrial uses. It is used in the manufacture of soap, paints, varnishes, waxes, seals, lubricants, inks, coatings, oilcloth, linoleum, plastics, rubber, drugs, chemicals and adhesives. etc.

India, Indonesia and Indo-Chinese countries are the chief producers of turpentine and rosin. An Indian non-coniferous tree, the Indian frankincense ***(Boswellia serrata)*** also yields turpentine.

Other turpentines of lesser importance are Canada balsam ***(Abies Balsamea)***, spruce gum ***(Picea rubens)***, Venetian turpentine ***(Larix decidua)***, Bordeaux turpentine ***(Pinus pinaster)*** and some others.

Balsams

They contain less oil than turpentines and are, therefore, thick, syrupy viscous liquids. They are highly aromatic due to the presence of **benzoic** or **cinnamic** acid. They find uses in medicines for dressing and healing of wounds and in coughs, colds bronchitis, etc. in the perfume industry as fixatives, as adhesives and flavourings, etc.

Some of the examples of balsams are balsam of Peru ***(Mytoxylon balsamum)***, balsam of Tolu ***(M. balsamum var. pereirae)***, styrax ***(Lievanti styrax)***, benzoin ***(Styrax spp.)***, copaiba ***(Capaifera spp.)***, etc.

Elemi

They are liquids at the time of exudation but harden on exposure. They comprise a number of different oleoresins. Manila elemi is obtained from the pili tree ***(Canarium luzonicum)***. African elemi is obtained from the ***Boswellia freriana***.

Elemis are used for torches, caulking boats, in the manufacture of adhesives, inks, cements, in varnishes, medicines and perfumes.

GUM RESINS

As stated earlier gum resins are mixtures of gums and resins. They exude as milky substance and harden into small irregular masses.

Ammoniacum is obtained from ***Dorema ammoniacum*** of **Umbelliferae**. It is used in perfumes and as a stimulant of the circulatory system.

Asafetida is obtained from ***Ferula astafoetida*** of **Umbelliferae**. It is greyish or reddish in colour. It has a foul smell and a bitter acrid taste due to the presence of sulphur. It is used for flavouring curries, sauces and other edible items. It is also used in medicines for cold, coughs, asthma, disgestive disorders and in perfumes.

Galbanum. It is obtained from ***Ferula qalbaniflua*** of **Umbelliftrae**. It has a powerful smell. It is used in medicines.

Myrrh. This gum resin is obtained from ***Commiphora spp.*** of the family **Burseraceae**. It is used in perfumes, incense, ointments, medicines, disinfectant and embalming mixtures.

Frankincense is obtained from ***Boswellia spp.*** of the **Burseraceae**. It is used as ceremonial incense in the Church, and in perfumes, face powders, pastilles, etc.

QUESTIONS

1. Name four gum yielding plants, Give their botanical names, and families.
2. Write notes on:

   (i) Canada balsam *(ii)* "Hing" *(iii)* Copals, and *(iv)* Amber

17

Industrial Plants (Contd.)

Sugars and Starches

Sugars and starches, the two common forms of carbohydrates are reserve food supply of not only plants but animals too. The two constitute necessary food for man. They are most valuable products of the plant world. The sugars are soluble but starches are insoluble carbohydrates. While the principle sources of sugars can be stem or underground roots, those of starches are cereal grains and certain underground tuberous root or stem structures. The principal sugar is sucrose.

The sources of storage sugars are;

(i) Roots–Beets, carrots, parnships, etc.

(ii) Stems-Sugarcane, maize, sugar maple, sorghum, etc.

(iii) Flowers—Palm.

(iv) Bulb–Onion.

(v) Fruits.

The sources of starches are *(i)* cereals like maize, wheat, rice, etc. *(ii)* potato, aroids; tapioca, arrowroot and sago.

In this chapter only sugarcane (sugar yielding) will be discussed. Sugar beet, sweet potato, tapioca and patato have already been described in Chapter 6.

Sugarcane

(Saccharum officinarum)

Origin and History

Sugarcane and sugarbeet are principal sources of sugar. They are sometimes called industrial plants because the entire sugar industry is dependent on them. Sugarcane is responsible for more than half the world's supply of sugar. The best production of sugarcane is in the tropics. Sugarbeet ***(Beta vulgaris)*** is cultivated in the cool temperate zone which is not suited for the sugarcane.

Sugarcane is believed to have originated in the tropics in or near **New Guinea** in the pre-Christian times (Fig. 17.2).

The plant has not been found in a wild state, and, therefore, its wild ancestor is not known. In 327 B.C. **Nearchus,** an officer in Alexander's army wrote that the barbarians beyond the Indus were able without the help of bees to make honey from the sap of a honey-bearing **reed**. On the basis of chromosome number ($2n = 80$) and morphology ***S. robustum*** and ***S. officinarum*** still occur in plenty in the New Guinea area today. *S. officinarum* (noble) canes is believed to have migrated to Asia. There it hybridized with the local ***S. spontaneum*** to produce ***S. sinense.*** These hybrid thin canes grow very well in the monsoon climates there. *S. officinarum* also spread to the Pacific Islands.

The plant was carried to Egypt in 641 A.D. and to Spain in *755* A.D. The Spanish and Portuguese explorers spread the plant to Madeira in 1420. By the end of the sixteenth century the crop was being cultivated in the West Indies and South America. It reached U.S.A. in 1741. Today sugarcane is widely cultivated in almost all the moist tropical and semi-tropical regions.

Sugarcane Growing Regions

WORLD. The chief sugarcane-producing regions lie between 30° north and 30° south of the Equator, *viz,* India, Cuba, Brazil, Philippines, Hawaii, Java, Puerto Rica, Australia and U.S.A. The first three countries are the major producers of sugarcane. Brazil is the largest producer of sugarcan. India is the second largest producer.

INDIA. In India, Uttar Pradesh and Bihar are by far the leading states in growing sugarcane. 4.63 hectares million acres of land are under sugarcane cultivation in India. The total yield is 322.94 MT.

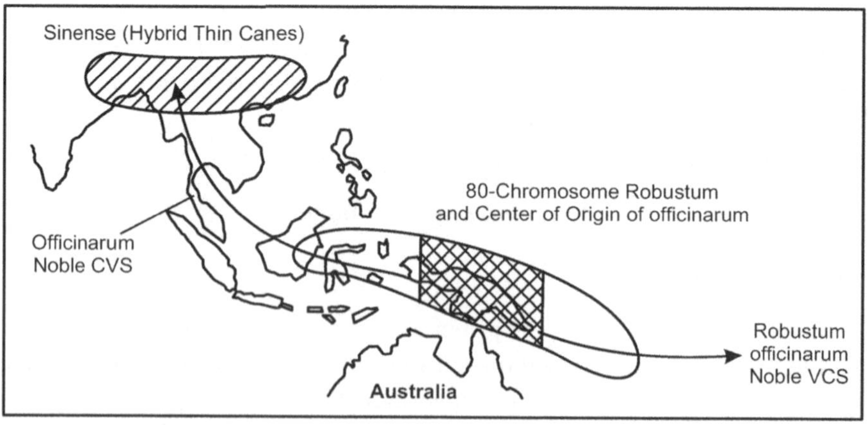

Fig. 17.1. Evolutionary distribution of sugarcanes.

Botanical Description

Sugarcane belongs to the family Poaleae. The plant is a coarse grass. It is quite tall. It has a height of 10–20 ft. A single plant may bear a number of stems in a tuft (Fig. 17.3). The stem is solid and jointed. The stems are supported by prop or stilt roots which arise from several root initials of the lower nodes. The nodes bear axillary buds. The stem bears a terminal inflorescence, but only few seeds are formed. The cultivated varieties of sugarcane are heterozygous and do not breed true. The plants are, therefore, propagated by vegetative means.

Fig. 17.2. Sugarcane plants.

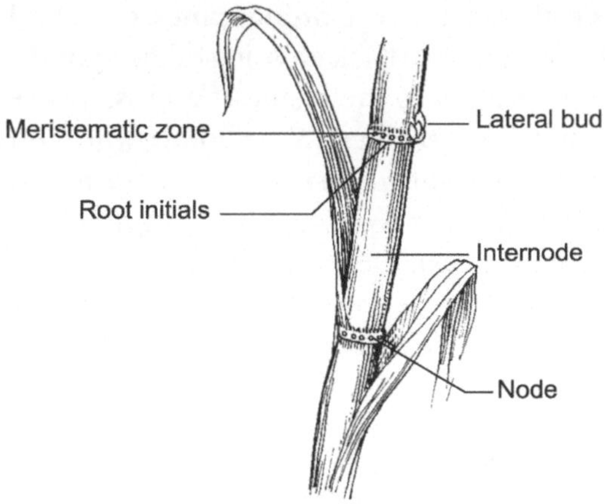

Fig. 17.3. Portion of the stem of a sugarcane showing lateral buds and a number of root initials.

Ecological Factors

Sugarcane is easy to grow. It needs a hot and moist climate (temperature averaging 70° F and an annual rainfall around 60 inches which may be supplemented by irrigation). Rich, deep and well drained soil gives best yield.

Cultivation

Modern method of cultivation is almost mechanical. In poor countries, however, cultivation may be done by hand at all stages. "**Sets**" cut from the upper part of old stems are usually used for cultivation. A 'set' or 'seed' has several nodes and internodes. The field is plowed deeply and 'sets' are placed in the furrows or trenches and are covered with soil. The soil is given an ample supply of nitrogen, potash and phosphate fertilizers. In about two weeks time the 'sets' sprout, i.e., the axillary buds from the upper nodes give rise to new shoots and the lower nodes produce rings of adventitious roots. A 'set' usually gives 2 or 3 crops called **ratoons**. It has, however, been found to produce aerial shoots (ratoon crops) continuously for as many as 20 years, under exceptionally favourable conditions. Utmost care is taken to promote profuse **tillering**. As soon as the tillers have developed their own roots, they are severed from the mother shoot and are carefully planted separately. Constant weeding is essential since sugarcane cannot withstand competition from weeds. Fertilizers should also be supplied to the field during the early stages.

Harvesting is done usually after one to one and a half year of the planting of the 'set'. Harvesting is usually done by hand by chopping the stem at the base with a cane knife. In advanced countries cutting machines are used for mechanical harvesting.

The important varieties of sugarcane are believed to have arisen as a result of crosses between the **soft sweet noble canes** and the **hardy wild canes** usually followed by one or more back crossing of the hybrid with these varieties. The existing superior variety is the result of a cross between the noble cane ***(S. officinarum)*** and wild species ***(S. spontaneum)*** commonly known as 'Kaab'. The latter is resistant to diseases, drought and other unfavourable conditions to which the former is susceptible. The hybrid produced as a result of this cross though hardier exhibit some bad qualities particularly the impurity of the juice. The hybrid is back crossed with the cultivated species, a process called '**nobilization**' for a number of generations till the desired combination of the characters is achieved.

Preparation of Sugar

Extraction of juice. The sugarcane stems are transported to sugar mill by road or rail. They are carried into the mill by big conveyors. The stems are cut into small pieces by crushers. They are then allowed to run through 3 sets of steel rollers. The juice is extracted in each expression. The cane pieces are wetted after each expression to facilitate complete extraction of the sugary sap. The dry fibrous residue is called **bagasse**.

Purification of juice. The dark-greyish sweet juice contains a number of impurities. It contains beside sucrose and glucose, small qualities of proteins, gums, organic acids, pectin, ash, minerals, dust particles, pigments and small pieces of cane. The soluble non-sugars are precipitated *(***clarification***)* and the insoluble substances are removed *(***defacation***)* to purify the juice. After filtering to remove the solid particles the juice is heated to coagulate the proteins. Lime is now added to neutralise the organic acids and also to prevent the conversion of sucrose into some inferior sugars. In addition to lime, chemical compounds like sulphur dioxide, sodium carbonate, carbon dioxide, phosphoric acid, etc., are added to the cane juice to precipitate out the non-sugars. The partly purified juice is then run through filter presses to remove the insoluble particles.

Concentration. The juice is boiled in vacuum pans to concentrate it into a thick syrup *("***massecuite***")* from which a solid mass of brown crystals separate out. The liquid remains called, **molasses** separated out through perforated bases or by using centrifuges. The molasses may be further evaporated to form sugar crystals or can be used as a cattle feed or can be fermented into rum, ethyl alcohol and vinegar.

Refining. The brown crystals (96% sucrose) are further refined to give the white crystals of sugar. The brown crystals are dissolved, washed, recrystallized and are decolourized with carbon. The refined sugar is usually packed in gunny bags for the purpose of transportation. Sugar is also packed in the form of powder and cubes.

Uses of Sugar

1. It is the chief sweetening material. It is extensively used all over the world for sweetening beverage, various kinds of food preparations, etc.
2. The bagasse is used as a fuel in the sugar mills. It is used as a source of paper and as an ingredient of fibre board.
3. The byproduct, molasses, is of great industrial value. It is used in the distillaries for the manufacture of rum, industrial alcohol and vinegar.
4. The molasses is used in cooking and candy making.
5. The molasses is a very good cattle feed.
6. The molasses is also sometimes used as a manure.

Other Sources of Sugars

Maple Sugar. The maple sugar is derived from the sweet sap of the stems of sugar malple *(Acer saccharum)* and the black maple *(A. nigrum)*.

Palm Sugar. Several species of palm are a very important source of sugars in the tropics. The sugar is derived from the tender parts of the stem (date palm) and flowers of wild date *(Phoenix sylvestris)*, the palmyra palm *(Borassus flabellifer)*, the coconut palm (*Cocos nucifera)*, the toddy palm *(Caryota urens)* and the gomuti palm *(Arenga pinnata)*.

Sorghum Syrup. It is derived from the stem of the sweet sorghum *(Sorghum vulgare* var. *saccharatum)*.

QUESTIONS

1. Describe the botany, method of cultivation and uses of sugarcane.
2. What is the botanical name of beet root? What is its economic importance?
3. Commercial starch is obtained from which plants. Write brief notes on two of them.

18

Fats and Oils

Vegetable fats and oils, also called fixed oils, are chemically esters of fatty acids and a trihydric alcohol, glycerol. Waxes are esters of fatty acid and a monohydric alcohol. Fats are solid or semisolid at ordinary temperature whereas oils remain in a liquid state. The fats contain **stearic** or **palmitic acid** and the oils contain **oleic acid.** They are actually colourless and tasteless but the colour and taste may develop due to the presence of traces of essential oils or pigments. If stored for long and at high temperatures they break into various aldehydes, ketones, etc., and develop **rancid taste** and smell. Edible fats and oils are so much in demand these days that the governments of various countries are paying great attention in finding out new sources of vegetable oils. In our country nine oilseed crops occupy 26 MH and the total Yield is 23.26 MT in 2006-2007.

Fats and oils are classified into four types:

(i) **Vegetable fats.** They are solid or semi-solid at ordinary temperature. They include coconut oil, palm oil, cocoa butter, nutmeg butter, etc.

(ii) **Drying oils.** They harden into elastic films on exposure to air. They include linseed oil, tung oil, walnut oil, perilla oil, soybean oil, safflower oil, etc.

(iii) **Semi-drying oils.** They form soft, solid films only after longer exposure to air. They include cottonseed oil, corn oil, sesame oil, sunflower oil, mustard oil, etc.

(iv) **Non-drying oils.** They do not form solid films on exposure to air. They include olive oil, castor oil, groundnut oil, etc.

The coconut oil, soybean oil, cottonseed oil, sesame oil, mustard oil and groundnut oil are more well known edible oils and are widely used as a cooking medium. Some of the oils are used for massaging the body and hairs. Oils are extensively used for manufacturing soaps and candles. They are also used in preparing paints, varnishes, furniture and leather polishes, oil papers, putty, laxatives, lubricants, cosmetics, waterproof fabrics, inks, artificial leather illuminants and oil cloth etc.

Extraction. The fatty oils are insoluble in water and are present in the form of insoluble droplets or deposits within cells. Most oils are stored in seeds, but they may also be found in fruits, stems, tubers, etc. They provide the much needed energy for the germination of seeds.

Extraction of oil involves separation of kernels from the seed coats, **hot** or **cold expression**, solvent or screw extraction and centrifuging. In hot expression the separated kernels are first cooked in steam to facilitate oil flow during extraction. In cold expression the kernels are directly crushed into a fine meal without any previous treatment. The oils are then removed from the meal by organic solvents in the case of non-edible oils and by means of screw or hydraulic pressure in the case of edible oils. The solvent evaporates at 60° to 113°C and the extracted oil is left behind. The pressure breaks the cell walls and the fat droplets are extracted.

The oil is then usually run over filter presses. The filtered oil may be purified further by heating (to coagulate albuminoids, etc.), or by treating with caustic soda (to saponify free fatty acids). The unsaturated edible 'oils' are now being converted into saturated 'fats' by the 'vanaspati' manufacturers since the latter is preferred very much by the modern housewife. The unsaturated oils (oleic acid, etc.), are hydrogenated by adding hydrogen under pressure to manufacture saturated fats (stearic acid, etc.), or the vegetable 'ghee'.

An account of the important oil-producing plants is given in this chapter. Soybean and groundnut have already been described in Chapter 4. Cotton is described in Chapter 12.

Vegetable Fats

COCONUT

(Cocos nucifera)

Origin and History

Coconut palm is a source of vegetable fat as well as the coir fibre. Coconut palm is widely distributed in the tropics. There are conflicting views about the center of origin of coconut. Coconut has been thought to originate either in South America, Polynesia or Asia. Most of the workers believe a southeastern Asian origin of coconut. It is supposed to have spread eastward and westward by means of water currents.

Coconut Growing Regions

It is grown in India, Sri Lanka, Indonesia, the Philippines, West Indies and in several islands of the Arabian sea, Bay of Bengal and Pacific Ocean.

The Philippines is the largest producer of coconut. India stands second in coconut production. In India coconut is grown extensively in the coastal regions of southern and eastern India, Tamil Nadu, Kerala, Maharashtra, Karnataka,

Gujarat and Bengal. In India coconut is grown in 1.95 MH and the production in 148.11 million nuts in 2006-2007.

Botanical Description

Cocos nucifera belongs to the family **Palmaceae**. It is a tall plant and attains a height of 80 ft. or more and lives up to 90 years but the dwarf varieties are small and live up to 30 to 35 years only. It has an unbranched stout trunk which has rings of leaf scars. The trunk has a swollen base which is surrounded by a mass of unbranched roots. The trunk bears at the tip a crown of very large pinnate leaves. The leaves are 6-18-ft. long and the leaflets are 2-3 ft. long Numerous male and few female flowers are born in spadix inflorescences in the axils of leaves when the plant is 8–10 years old. Single seeded, three angled fruits are borne in clusters at the top of the plant (Figs. 18.1). They may be 6"–12" in length. The colour of the fruit varies from dark green to deep orange or brick red. The fruit is a drupe. The exocarp and mesocarp are thick and **fibrous** (source of coir) whereas the endocarp (shell) is a hard structure. The shell 'eyes' bears three 'pores' at one end. The embryo lies below one of these pores. Internally there is thick albuminous endosperm (meat) attached all over the endocarp. The endocarp encloses a large cavity which is filled with watery fluid in the young fruits.

Fig. 18.1. Coconut. A. Plant; B. Cluster of fruits; C. A drupe split to show endosperm.

Ecological Factors

Coconut palm is commercially grown in the hot and humid tropics from 20°N to 20°S of the equator. It requires 50" to 90" of well distributed rainfall and an uniformly high temperature around 80°F. The plant neither grows well nor produces fruits in places which get less than 40" of annual rainfall or temperature below 70°F. Well drained alluvial or sandy soils of sea shores, river banks and hill slopes are very good for its growth.

Cultivation

The tall varieties are more important for commercial cultivation in large plantations. The dwarf varieties, which start flowering and fruiting only four years after the plants are planted are grown only to a limited extent.

Mature fruits (12 months old) are selected from high-yielding varieties (100 nuts per annum) which are healthy, middle aged and have other desirable characters. The seedlings are carefully raised in nurseries in well drained seed beds. The shoot appears after 4 months. About 50% of the seedlings are selected when they are 7 to 8 months old.

The healthy seedlings are transplanted in pits which are usually 3' × 3' × 3'. The pits are dug up 25 to 30 ft. apart about 1 to 3 months earlier and are manured and disinfected before transplanting them. The plant requires constant manuring. Wood ash (obtained by burning leaves, husks etc.), cattle manure and occasionally green manure are usually given during the rainy season. Potash is very much needed for giving higher yields.

The nuts are harvested 6 to 8 times in a year.

Extraction of Oil

The harvested nuts are cleared off the husks. The shell is split open by hand or by machine. The dried 'meat' is called **copra**. The copra or sometimes fresh 'meat' is crushed before oil is expressed. The copra is usually **hot-expressed** to yield pale yellow or colourless coconut oil. The residue may be run through hydraulic presses to remove more oil. The total yield of oil is as high as 70 per cent. The coconut oil is then refined.

Uses of Coconut

1. Coconut oil is very widely used as fatty oil. Refined coconut oil is used in preparing various food products and confectionery.
2. Coconut oil is used as a hair oil. It is used in the manufacture of quality soaps, shampoos, shaving cream, cosmetic, etc. Glycerin is an important byproduct in the soap manufacture process.

3. The oil is also used as a lubricant, fuel, etc.
4. The press cake is used as a stock feed.
5. The fresh "meat" is eaten raw and is also used in preparing sweetmeats.
6. The husk particularly the green husk is the source of the fiber-coir, which has several uses.
7. The trunk of the tree is used like a girder in the construction of huts, etc. The leaves are used for thatching purposes and as fuel.
8. In many tropical countries the sap of the coconut palm is tapped by cutting the end of the unopened spathe surrounding the young inflorescence. The sap is used as a drink called 'toddy'. The sap contains a rich amount of sugar.

OIL PALM

(Elaeis guineensis)

The oil palm is indigenous to hot and humid tropical zone of **West African countries like Zambia, Liberia, Sierra Leone, Nigeria, Guinea, Congo, Ivory Coasts, Togoland, Cameroon etc**. Today demand for palm oil has increased so much that it is being cultivated on a commercial scale in Malaysia, Indonesia, West Indies, and Mexico.

Botanical Description

The oil palm tree, (***Elaeis guineensis***) belongs to the family **Arecaceae.** The plant has a stocky stem with a crown of 20 to 40 very large leaves at the top. The tree may grow upto a height of 50 to 60 feet. The thick stem is covered with leaf bases of the fallen leaves as seen in coconut palm. The compound leaves have 100 to 160 pairs of leaflets (Fig. 18.2). The flowers are unisexual and the plant is monoecious. The lower leaves bear spadix of female flowers and the upper leaves bear spadix of male flowers. The tree flowers when they are 4-6 years old. The spadix or the compound spike has numerous spikelets arranged in a spiral manner.

The tree bears as many as 200-2000 fruits in a bunch. They are formed twice in a year in the rainy season as well as in the dry season till the plant is about 40 years old after which the fruits are borne in lesser number for a few years more. The fruits are drupes.

Harvesting and Extraction

The fruits are brought down by professional climbers who cut down the bunches of fruits with the help of a sickle. The fruits are boiled in water or are dried in the sun. There are two types of oils obtained from the plant: palm oil from the **pulp (mesocarp)** and palm kernel oil from the **endosperm**. The fibrous pulp of the fruit is separated from the nut by pounding the fruit by wooden sticks. The disintegrated pulp can yield about 55% of the oil content of the fruit. The separation of the pulp can also be achieved by the process of fermentation. The pulp is kept overnight when the palm oil is released and stored in a container. The oil is then boiled and clarified.

Fig. 18.2. Oil palm tree.

Palm Kernel Oil accounts for the remaining 50% of the total oil content of the fruit. The kernel is actually the endosperm present within the stony shell i.e. the endocarp of the drupe. The shell is broken by hammering with a piece of stone. The kernels can either be locally subjected to crushing for extraction of oil or are exported to European countries or U.S.A. for extraction. The kernels are subjected to mechanical presses for extracting the oil.

Uses of Palm Oil

1. Palm oil is widely used as a cooking medium.
2. Palm oil is a source of vitamin A.
3. Palm oil is used for manufacture of candles, soaps, and lubricants.
4. Palm kernel oil is used for the manufacture of shampoos, candles, soaps, glycerine etc.
5. Palm kernel oil is used in bakeries for preparing cakes, biscuits, butter like preparations etc.
6. The left-over cake is used as cattle feed.

Drying Oils

LINSEED

(Flax; Vern. Alsi)

(Linum usitatissimum)

Origin and History

Linseed of flax is a source of an oil and a fiber. According to **Vavilov** the small seeded types of linseed originated in **southwestern Asia (Afghanistan, etc.)**. It spread northward to Europe and other parts of Asia and southward to India. The northern forms evolved into early races with long unbranched stems whereas the southern forms evolved into late races with lot of branching in the stem.

Linseed Growing Regions

Linseed as a source of fiber has been in use of more than 4,000 years. It is now extensively grown in Canada, United States, Uruguay, Argentina, U.S.S.R., India and China.

In India the chief linseed growing areas are Bihar, Uttar Pradesh, Madhya Pradesh and Andhra Pradesh. It is also cultivated in Bengal, Punjab and Maharashtra.

Botanical Description

Linseed belongs to the family **Linaceae**. It is a herbaceous annual herb reaching a height of 1 to 4 feet. The Indian linseed is of a shorter height of about $2^{1/2}$ feet. The oil-yielding variety of the plant is branched which, therefore, results in large number of seeds. The leaves are small, narrow and blunt at the apex. The inflorescence is a terminal panicle of small white or blue flowers. The fruit is a capsule which is commonly known as the 'seed ball'. The seeds are quite small and flattened. They are white or dark brown in colour.

The seeds of linseed have a very high oil content (32 to 43 per cent.) The predominant oil components are **linolenic**, **oleic** and **linolic acids**.

Ecological Factors

Linseed grows in temperate zone but is also grown in tropical climates for the purpose of oil. It thus grows in different types of climates. It is grown as cold season *(**rabi**)* crop in India. When grown for fiber, it is grown in cool, moist summer climates of the temperates.

Linseed grows in almost all types of soil, where sufficient moisture is available. Good porous soils give the best results. In India linseed is of two types : **a peninsular type** and an **alluvium type**. The peninsular type is grown in the black cotton soil, to which it is especially adapted due to its deep rooting habit, quick growth and early maturity, which are all desirable characters on account of the cracking of the soil and the consequent rapid loss of moisture. The alluvium type which is shallow-rooted, slow growing and late ripening is more adapted to the alluvial soils of the Indo-Gangetic plains.

Cultivation

A fine, firm and porous seed bed is prepared by ploughing the land several times, followed by harrowing. The land is then rolled or planked. The seeds are sown usually by means of a drill or by dropping seeds through a tube attached to a plough. Sometimes the seeds are sown broadcast. It is sometimes sown as a mixed crop with gram or 'rai' ***(Brassica juncea).*** When sown alone in lines, the lines are a foot apart.

The crop is usually harvested in 3 to 4 months time after sowing. Harvesting is done by hand as well as by machine. The plants are either uprooted or are cut by sickles. They are then taken to the threshing floor where the seeds are beaten out and winnowed.

The seeds are cleaned and crushed to a fine meal. Oil is then extracted by heating or by means of solvents. Cold expression is practised when the oil is used for edible purposes. The oil is heated to 125°C to increase its drying property. Linseed oil is a very important **drying oil**. It is one of the oldest of oils and has been used as a drying oil for centuries.

Linseed oil is yellow-brown. The liquid oil can be kept almost indefinitely in closed containers, but it sets to a hard, elastic film when exposed to air.

Uses of Linseed Oil

1. The oil is chiefly used in the manufacture of paints, varnishes, linoleum, oil cloth, soft soap and printing ink.
2. The refined and unheated oil is used for cooking purposes in Russia. and countries of Eastern Europe.

3. The press cake is used as stock feed provided the hot pressure process has been used. The cold-dressed cake is poisonous due to the presence of a cyanogenetic glucoside, **Limarin** The heating prevents the evolution of **prussic acid** from the glucoside and so it becomes edible.
4. The oil is used as fuel and in leather production.
5. The flax fiber got from the stem has a variety of uses which are discussed in Chapter 12.

SAFFLOWER

(Carthamus tinctorius)

Origin and History

Safflower oil is **a drying oil** obtained from the seeds of ***Carthamus tinctorius*** of the family **Asteraceae**. The plant is also well known as a source of an orange-red dye. It is **native to India**. It is widely cultivated in India, the West Asian countries, Egypt and many other east African countries, China and to some extent in U.S A. and southern Europe. Safflower seeds have been found in some 3,500 years old Egyptain tombs.

According to some safflower is indigenous to mountaneous regions of **Abyssinia** and **Afghanistan**.

In India safflower is cultivated as an oil seed crop in Maharashtra, Madhya Pradesh, Andhra Pradesh, Tamilnadu, and Karnataka. It is grown for dye in northern India.

Botanical Description

The plant is an annual herb. It grows very rapidly. It is highly branched. It grows to a height of 1 to 4 feet. Leaves are crowded near the base of the stem in the form of a rosette. Leaves are usually arranged in a spiral manner and are borne at irregular intervals. The leaves occur very closely at the tips of the branches and form involucral bracts of the inflorescence. The 'head' inflorescence is formed at the tip of the each branch of the plant. All the florets of the inflorescence are regular and tubular. The fruit is an achene.

The seed contains 24 to 36 per cent of a very good drying oil. It does not dry as rapidly as linseed oil.

Cultivation

Safflower is a **rabi** crop. It is often grown as a mixed crop along with wheat and jowar. The seeds are sown in rows alternating with the main crop. The apical portion of the plant is removed after 2 months so that the plant becomes bushy so as to produce large number of flowers and fruits.

Harvesting and Extraction

Four month old plants are uprooted and the harvested crop is allowed to remain in the field in the form of heaps. The plants are beaten with sticks to remove the seeds from the fruits.

Safflower oil contains several fatty acids such as palmitic acid, stearic acid, arachidic acid, oleic acid, and linolic acid. Linolic acid and olelic acids are the principal components of the oil.

The oil is extracted by crushers or by hot dry distillation. The other methods employed for extracting the oils is by hydraulic process or solvent extraction process.

If the crop is cultivated for extracting the dye the flowers are picked up and the petals are carefully removed. The petals contain reddish carthamin, yellow coloured safflower yellow and isocarthamin. Carthamin is soluble in alkalies but not in water whereas the yellow pigment is soluble in water.

Uses of Safflower Oil

1. In India the oil is mainly used for food and cooking.
2. In other countries it is chiefly used in the paint and varnish industry.
3. Decorticated (without fibrous seed coat) seed cake is a good stock feed (20 to 55% protein).
4. The dye is used for dyeing cotton, silk and wool fabrics. It gives various shades of orange, red and yellow colour to the cloth.
5. The oil is used in industries for the manufacture of soaps.
6. The seeds have medicinal value as a tonic and diuretic.

It is possible to grow three sunflower crops a year with an average yield of 20 to 25 quintals per hectare per season. On account of its day-neutrality, sunflower crop can be adjusted in any crop rotation and may be helpful in solving the problem of non-utilisation of land in off-season. This necessity has been increasingly felt with the extension of irrigation facilites.

Semi-drying Oils

SESAME

(Vern. Til)

(Sesamum indicum syn S. orientale)

Origin and History

Sesame is believed to be indigenous to **India** though some workers think Euphrates valley of Iraq or Africa to be the original sites of origin. Archaeological findings of sesame from Palestine and Syria have been dated to 3000 B.C. The crop is, however, believed to be **African** since two-thirds of the species occur in tropical Africa On the other hand no wild species have been found in West Asia.

India is the largest sesame-growing country. It is also an important crop in Burma. The crop is grown in India from ancient times. Though most of the workers do not believe in its Indian origin. **Nayar** and **Mehta** (1970) believe it to have originated in **Ethiopia** or **peninsular India** or in both independently. Some carbonized seeds have been found from the ancient site of Harappa, where Indus valley civilization flourished about 5,500 years ago. 'Til' seeds are very commonly used in the religious rites of the Hindus indicating its domestication in the remote past. It is also believed to have been under cultivation in China from the ancient times.

Sesame Growing Regions

The chief sesame-growing countries lie in the tropics as well as in the temperates. They are China, India, Burma, African (Sudan, Nigeria, Uganda) and Latin American countries (particularly, Mexico). Sesame oil is a very important vegetable oil in China, India and Mexico. In India sesame is cultivated in Uttar Pradesh, Madhya Pradesh, Rajasthan, Maharashtra, Andhra Pradesh and Tamil Nadu.

Fig. 18. 3. Top of ***S. orientale*** showing capsules.

India occupies the first position in the world in the production of sesame (about 25% of total yield). The crop occupies about 40% of the total cropped area under sesame in the world.

In India sesame is grown either alone or as a mixed crop along with maize, jowar, bajara, or cotton etc.

Botanical Description

Sesame belongs to the family **Pedaliaceae**. The oil is also known as **benne** and **gingelly**. ***Sesamum*** has about 36 species. The plant is an annual bushy herb (Fig. 18.2). The leaves bear 1 to 3 axillary flowers. The fruit is a capsule. Dehiscence is by means of apical pores. The seeds are white, brown or black in colour. The seeds have about 50% oil (oleic and linolic glycerides) 18 to 25% carbohydrates and 20 to 24% protein.

Cultivation

It grows well in warm climates of the tropics and subtropics. It requires plenty of water (50 to 110 cm annual rainfall) though it is a drought resistant crop. Sandy soil is good for its cultivation. The seeds are sown broadcast or in rows. It is rainfed **kharif** crop

The plants are harvested by hand before they dry up. The plants are cut and dried in heaps. The seeds are removed from the capsules by shaking the cut plants or by beating or trampling upon them.

The sesame oil is extracted by cold pressure. The oil is of **semi-drying type**. The presence of two unique compounds **sesamin** and **sesamolin**, which are powerful **anti-oxidants** does not allow it to turn rancid.

Uses of Sesame Oil

1. Sesame oil is used as a cooking medium. It is the principal vegetable oil in Mexico and India. It is often hydrogenated to produce vegetable 'ghee'.
2. It is used in making margarine, shortenings and other food products.
3. The poorer quality of the oil is used in the production of soap, perfumery, rubber substitutes, hair oil and paints.
4. The seeds are used in the confectionery and baking industries, 'Laddoo', 'rewri', and 'gazak' are prepared from the seeds.
5. In Africa sesame forms a part of diet in the form of porridge, soup and sweets.
6. It is also used in medicine.
7. It is rich in vitamin E.
8. Sesame oil is a very good substitute for olive oil.
9. Sesame oil is used as a medium for steroid hormones, antibiotics vitamins etc.
10. The oil is used along with the insecticide pyrethrum.
11. Sesame oil cake is used as animal feed.
12. The roasted seeds are powdered for making cakes.

SUNFLOWER

(Helianthus annus)

Origin and History

Sunflower oil is another example of semi-drying oil. It is obtained from seeds of ***Helianthus annus*** of the family **Asteraceae**. The plant is a native of **Peru** or **Mexico** of Central America (Fig. 18.4).

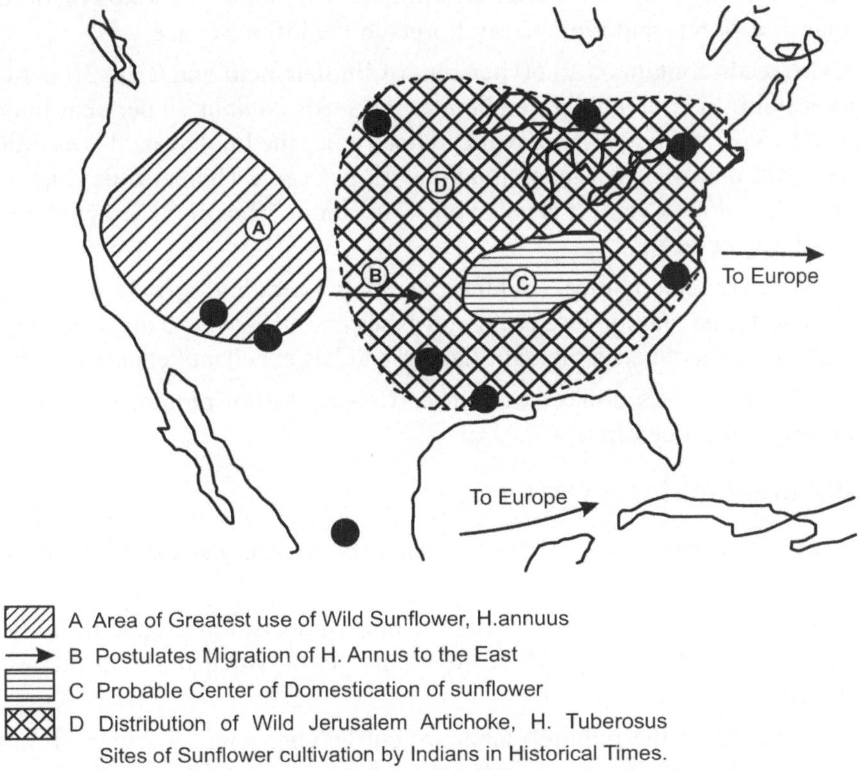

Fig. 18.4. Evolutionary geography of sunflower.

Sunflower is extensively cultivated for oil in the temperate regions of Russia and European countries like Spain, Hungary., Rumania, Bulgaria, and Argentina and to some extent in U.S.A., Canada and the tropical parts of India, Asia and many of the African countries. Russia is the largest producer of sunflower oil. In India it is grown in 2.13 MH and the yield is 1.12 MH.

In the present time sunflower seed oil is playing an important role in the world's vegetable oil industry. It has come to acquire the third position among the edible oils and soap trade in the international markets. Until 25 years back it occupied the ninth position.

In India sunflower cultivation has picked up in the recent years. It is being cultivated on a large scale in Punjab, Haryana, Karnataka and Maharashtra.

Botanical Description

The plant is a coarse, tall, annual. It has a woody stem which is covered with rough hairs. It is of 5 to 10 feet in height. Large, ovate or cordate, dark green coloured leaves are borne in alternate manner. The inflorescence is capitulum or head which is usually terminal. There are two kinds of florets– tubular disc florets and ligulate ray florets in the inflorescence.

The seeds contain 55 to 60 per cent of **linoleic acid** and 25 to 30 per cent of **oleic acid**. In cool temperate climates the seeds contain 70 per cent linoleic acid and 15 per cent oleic acid, while in hot tropics the linoleic acid is as little as 20 per cent and the oleic is as much as 65 per cent. The **semi-drying oil** is usually 32 to 45 per cent. It is pale yellow in colour. The oil is extracted by **hot or cold expression**.

Sunflower seed oil meal contains over 40 per cent **high grade protein**. It is sweet in taste and quality wise is as good as sesame and soybean meal. Sunflower oil after refinement gets a pleasant flavour and has excellent keeping quality.

Sunflower seeds unlike soybean and copra do not require sophisticated machinery for processing.

Cultivation and Harvesting

Several varieties of sunflower particularly the Russian varieties are today widely grown in India.

The seeds are sown in rows with the help of drills. Large spacing between the plants ensures better yield. If the plants get crowded thinning process is practiced.

The stalks of the mature plants are cut leaving about 3 inches from the base. The stalks are left in the open for drying them. The capitulum inflorescences of the plants are separated for threshing purposes.

Uses of Sunflower Oil

1. The oil is used as a cooking medium. It is a very good salad oil and is used in margarines. It is also used in the manufacture of butter substitutes.
2. The semi-drying oil is used in the manufacture of paints, varnish, soaps and cosmetics.
3. The seeds are a good poultry food.
4. The cake is used as a stock feed.
5. Sunflowers yield honey and wax.

MUSTARD

(Vern. Sarsoan, Rai, etc.)

(Brassica spp.)

Origin and History

Mustard, rape or colza oils are produced from the seeds of several species of **Brassica**. Some of them are: ***B. campestris*** var. ***sarsoan*** (yellow sarsoan or Indian colza), ***B. campestris*** var. ***dichotoma*** (black sarsoan), ***B. campestris*** var. ***toria*** (Tora or Tori), ***B.juncea*** (Rai), ***B. napus*** (the rape), ***B. rapa*** and some others. The origin of mustards has been described in Chapter 8.

They are now grown extensively in China, India, Japan and Europe. In India they are cultivated in Punjab, Haryana, UP Bihar, Bengal and Assam. In our country the mustard oil is normally derived from a mixture of yellow sarsoan, black sarsoan, 'toria' and 'rai'. There is often a mixed cultivation of these plants. In India mustard is cultivated in about 7 MH and the yield is about 7 MT.

Botanical Description

Brassica belongs to the family **Brassicaceae**. The plant is a sparsely branched annual herb. It is often cultivated mixed with wheat or barely during the winter season. It grows well in a medium loam soil. The plant bears beautiful yellow flowers in a terminal raceme. The fruits are long siliqua. The seeds are small, rounded and yellow or black coloured.

The seeds contain 30 to 45 per cent oil. The oil consists largely of glycerides of erucic, **oleic** and **linolic acids**.

Oil Extraction

On a small scale, in the villages, oil is extracted by crushing the seeds in wooden crushers ('Kolhu') drawn by bullocks. On a large scale the oil is extracted in large oil mills. During crushing 2 to 4% of water is added and the oil is pressed out ($1\frac{1}{2}$ to 2 hrs.) at a temperature of 31 to 40°C. The glucoside contents are completely hydrolysed under these conditions.

The crude oil is yellow, light brown or greenish in colour with a characteristic sharp taste and pungent smell. On purification with sulphuric acid or caustic soda, the crude oil gives rise to a pale yellow oil of high viscosity, with properties midway between those of the **semi-drying** and **non-drying oils**.

BARC has developed TM-1 as a result of mutation breeding of mustard. It has a higher oil content than the popular Rai-5.

Uses of Mustard

1. Mustard oil is edible and is the chief cooking medium for vegetables is northern part of our country. It is used for frying a variety of food preparations.
2. The oil is also used as fuel in lamps, etc.
3. The oil is used in the manufacture of soap and rubber substitutes.
4. Mustard oil is used as a lubricant in machineries, etc.
5. The oil is used for massaging the body and as hair oil.
6. Mustard seeds particularly rai are used as spice in vegetable and pickle preparations.
7. With camphor it is used for giving relief from muscular rheumatism and stiff neck, etc.
8. The oil cake known as '**Khali**' is very nutritious and is used as a cattle feed.

Non-drying Oil

CASTOR

(Vern. Arandi, Reri)

(Ricinus communis)

R. communis seeds are the source of a very important industrial **non-drying** oil, the castor oil. Castor belongs to the family **Euphorbiaceae**. The plant is probably native of **Africa**. Now the plant grows in wild as well as cultivated state in the whole world in both the temperate and tropical regions. It is chiefly cultivated in Brazil, India, Mexico, Russia. and Manchuria.

The plant is a coarse and erect herbaceous annual shrub. The plant grows to a height of several feet. It has a well developed tap root system. The stem is branched either all over its length or only in the upper part. The leaves are large and palmately lobed. The plant bears a terminal inflorescence and has unisexnal flowers. The fruits are spiny (Fig. 18.5).

The fruit is globular and has three locules each containing a single seed.

New varieties of castor have been bred to facilitate mechanical harvesting of the fruits.

The flowers are very small and fruits are spiny (Fig. 18.6). The fruits contain large mottled seeds. The seeds contain 35 to 55 per cent oil.

The oil is extracted from the seeds by cold expression. Oil may be further extracted from the press cake by means of organic solvents. The oil is thick, colourless or greenish.

Uses of Castor

1. The castor oil is of medicinal value. It acts as a purgative.
2. The oil has a number of industrial uses. It is used in the manufacture of soaps, inks, plastics, imitation leather, linoleum, oil cloth, nylon, etc.
3. The oil is a very good lubricant. As a lubricant it is used in a number of machineries including that of aeroplane.
4. The oil is used in a number of articles because of its water resistant quality. It is used for preserving leather.
5. Castor oil becomes a **drying oil** if it is heated and dehydrated. It can then be mixed with more expensive oils in the paint industry.

Fig. 18.5. *Ricinus communis,* Leaves, flowers and fruits.

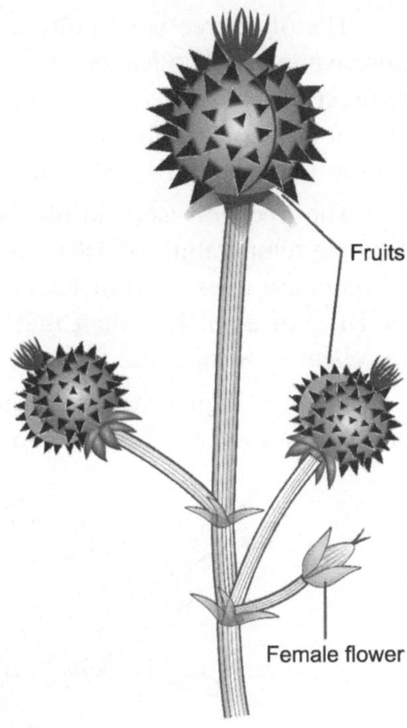

Fig. 18.6. *Casteor fruits*

6. Sulfonated castor oils are used as a dyeing aid and in finishing fabrics and leather.
7. The oil is also used as a fuel in lamps, etc.
8. The oil cake is poisonous but is a good fertilizer.
9. The stem is a source of paper pulp and cellulose.

OLIVE OIL

(Olea europaea)

Although the olive tree originated in West Asia, it has been cultivated for over 3000 years in the Mediterranean countries, where much of the olive crop is used to extract olive oil from their fruits. The chief olive producing countries are Italy, Spain, Greece, Turkey, Tunisia, Morocco, Portugal, Algeria and Libya. Olive tree has not been successfully grown in India.

Botanical Description

The olive tree belongs to the family **Oleaceae.** Hundreds of cultivars have been developed for obtaining oil from their fruits.

The olive tree is xerophytic in habit. It attains a height of 35 ft to 60 ft. approximately. The leaves of the tree are very narrow. The plant bears a large number of one seeded drupes (Fig 18.7). The mesocarp or pulp of the fruit is rich in oil and accounts for about 75% of the total oil output. The seed or the **'stone'** contains 12%-28% oil.

The tree flourishes in places which have warm and dry summer having average temperature of 18°C. The rainfall is 60-75 cm per year. The winter temperature does not fall lower than—5°C. The average winter temperature of 10°C or even less than that for a period of about 2 months is ideal for excellent flowering and fruiting. They grow best on calcareous sandy loams.

The tree is propagated by cuttings or grafting. The tree keeps on bearing fruits from the age of 5 years to the age of 50 years or more fig. 18.7.

Fig. 18.7. A twing of olive tree.

Extraction

The fully ripe olive fruits rich in oil are pitted and ground to a thick pulp. The pulp is then pressed to remove the juices, which are placed in a centrifuge to separate water from the oil. The pulpy matter extract along with the kernels are generally expressed several times. The cold-pressed product of the first pressing of olives is valued for intense flavour and low acidity. This is known as **'extra virgin'**. This oil can range in colour from a clear pale yellow to golden-green to bright green.

The olive oil contains 65–86% of oleic acid, 7–20% of palmitic acid, 5–15% of linolenic acid and very small amounts of myristic and stearic acid.

Uses of Olive Oil

1. A tablespoon of olive oil contains 14 grams of fat and 120 calories. The fat is mostly unsaturated (77%) which has a beneficial effect on blood cholesterol.
2. Olive oil is widely used as a cooking medium.
3. It is used as salad oils and for canning sardines etc.
4. It is used in the manufacture of expensive toiletries and cosmetics.
5. Green fruits of olives are pickled in brine so that the bitter glycoside (oleuropein) is neutralised.
6. The oil is used in textile industry.
7. The oil is used for medicinal purposes. It stimulates digestion and relieves stomach upset, flatulence and heart burn. It has vitamin E which is an antioxidant.
8. The residue of olive fruits after extraction of oil is used as cattlefeed and also as manure.

JATROPHA

Origin and History

Jatropha, the wonder plant, is native to **Mexico** and **Central America**. It originated in the **Caribbean** from where it was carried as a valuable hedge plant to Africa and Asia by Portuguese traders. It grows in many tropical and sub-tropical areas including India, Africa (particularly Egypt) and North America. In India it is grown in many parts.

Botanical Description

Jatropha has 175 species. It belongs to the family **Euphorbiaceae**. The plants are succulent shrubs and trees. ***J. Cureas*** is a deciduous species. It has a smooth gray bark.

The trees are deciduous. The leaves fall down during the dry season. The mature trees are small generally 3 to 5 meters in height. If conditions are favourable it can attain a height of 10 meters. It has large green to pale-green leaves. They have either alternate or sub-opposite phyllotaxy. The leaves are spirally arranged on the stem. They are 3 to 5 lobed. The petiole in 6 – 23 mm in length. The leaves bear axillary inflorescence which are either male or female. Flowering occurs during the wet season (May to September). There are two flowering peaks. If the atmosphere is permanently humid flowering occurs throughout the year. The flowers are pollinated by insects like honey bees.

The fruits are produced in winter season (July to November) when the shrub is leafless. However the plant can produce several crops during a year if soil moisture is good and the temperature is high. An inflorescence generally produces 10 or more spherical or oblong fruits. When the seeds are mature three bi-valved cocci are formed. The seeds mature three months after flowering.

Ecological Factors

Jatropha is a perennial plant. It can grow in arid conditions (even deserts) on any kind of ground. It is drought as well as pest resistant. The plants can be easily cultivated even in marginal poor soils with low nutrient content.

Jatropha is a fast growing plant. It is highly desirable in the sense that it not only prevents soil erosion but also increases the fertility of the soil.

Jatropha can be intercropped with other cash crops such as coffee, sugar, fruits and vegetables. ***Jatropha*** is moot successful in drier regions of the tropics with annual rainfall of 300 – 1000 mm. It occurs mainly at lower altitudes (0 – 500 m) in areas having temperature above 20°C. It can, however, tolerate even slight frost. In prefers well-drained soils having good aeration.

Cultivation

Jatropha has become a very important **bio-diesel** or **bio-fuel** crop. As stated earlier they grow even on marginal, saline, acidic, alkaline soils and slopy lands. They don't require irrigational facilities, and can flourish even in dry-land farming. It likes heat and is highly adaptable.

Extraction of Biodiesel

Jatropha seeds have an oil content of 37%. The oil contains 21% saturated fatty acids and 79% unsaturated fatty acids. The seeds are not edible by human beings or by grazers due to presence of a toxic chemical. The crop has fifty years of life. The first harvesting of the fruits is done after 2 to 5 years depending upon the soil quality and rainfall.

The crop produces 1500 to 10,000 kilograms of seeds per hectare and yields 2 to 4 tones of **bio-diesel** after esterification.

Uses of Jatropha

1. The oil when burnt even without being refined gives a clear smoke free flame. The byproduct is press cake.
2. The non-edible vegetable oil obtained from the plant is an alternative to diesel oil since it bears oil bearing nuts.
3. It is widely used for making soaps.
4. It is of medicinal importance. The latex of ***Jatropha*** has an alkaloid known as "**jatrophine**" which is anticancerous. It is used for treating paralysis, dropsy etc.

   Roots are antidote for snake-bites. The juice of the leaf is used for treating piles. The twigs of the plants are used for brushing teeth.
5. It is useful in treating domestic livestock for skin diseases, rheumatism and sores.
6. A blue coloured dye obtained from the bark of the trees is used for colouring cloth, fishing nets etc.
7. The leaves are used as food for the tusser silkworm.
8. ***Jatropha*** oil cake is rich in nitrogen, potassium and phosphorous. It can be used as organic manure.

The Government of India has planned to plant 30 million hectares of waste-land with ***Jatropha*** to produce 60 million tones of bio-diesel by 2030.

QUESTIONS

1. Name the chief oil yielding crop of India. Describe its morphology, cultivation and uses.
2. Describe the botanical features of a vegetable oil yielding plant.
3. Name the palm oil yielding plant. Give its botanical name, origin and distribution.
4. Why has **Jatropha** become so useful in our country? Give its botany and the climatic conditions for its cultivation.

19

Essential Oils

Introduction. Several thousand species of plants belonging to about 60 families possess a group of highly aromatic volatile organic substances called 'essential oils'. These differ from the fixed fatty oils of the plants in several ways. Unlike the fixed plant oils they do not give any oily feel and evaporate when they come in contact with air. Chemically they are **terpenes** which act as carriers of the **aromatic substances**. Terpenes are hydrocarbon in nature, being made up of isoprene units (C_6H_8), and existing as unsaturated straight chain molecules, or as ring structures, readily combining with other organic groupings. Most essential oils also contain camphors, and the more odoriferous compounds present in them consist of oxygen derivatives of terpenes, alcohols, esters, aldehydes and ketones. Besides terpenes, essential oils may also be oxygenated and sulphuretted oils. The essential oil have smaller molecules, usually less than 20 carbon atoms long. The term 'essential oil' was used during the middle ages because the pleasant smelling, highly volatile liquids were considered to be essential constituents of the plants.

Nature. The manner in which the essential oils are formed in the plants is not clearly understood. They are believed to be by products of metabolism and are found to be greatest in regions of photosynthetic activity. They are secreted in the plants either into external glands or epidermal hairs outside the main plant body' or into internal glands which may develop in any part of the plant tissue, flowers (rose), fruits (orange), leaves (mint), bark (cinnamon) rhizome (ginger), wood or seeds (cardamom), etc. Secreted oil is deposited in the intercellular spaces where an oil sac is gradually developed due to the breaking up of the cell walls. Chemically the oils may be different in different parts of the plant.

Function. The exact function of the essential oil is not known. They are believed to perform a number of functions, attract insects for pollination, repel animals and parasites, act as wound fluids; minimise the effect of heat on transpiration, act as hydrogen donors in oxidoreduction reactions, act as potential sources of energy and act as an antiseptic substance.

Uses. Essential oils find a number of uses. The most important use of the oils is in the manufacture of perfumes. They are used in perfuming soaps, deodorants and toilet preparations. The soils are used for flavouring various food and beverage products and tobacco. Essential oils are also used as stimulants and antiseptics, as constituents of medicines, as a laboratory reagent, as solvents in the paint industry, as insecticides, and as a component of pastes, polishes, ink, glue, etc.

EXTRACTION OF ESSENTIAL OILS

There are four methods for extracting the essential oils from the solid portion of the plant. These are, **distillation**, **expression**, **cold fat extraction** and **solvent expression**.

1. **Distillation.** This method is employed most widely to extract the essential oils from various plant materials. In case of flowers with delicate smell this is, however, not usually practised since heat may destroy the oils or may bring about certain chemical changes in them. Distillation in the case of immiscible oil is brought about with boiling water or steam. Boiling water or preferably steam is allowed to pass over the crushed material kept in a still when the essential oil is freed from the oil glands. The oil is vapourized along with water and the two are cooled and condensed in an adjoining condenser. The oil or **essence** separates from water and either forms an upper or lower layer depending upon it's density. The oil is removed and filtered. This oil may be a mixture of essential oils. They can be separated or purified further by fractionation technique described below.

Distillation in the case of miscible oil is by fractionation or rectification techniques. Plant materials or mixtures of essential oils obtained from steam distillation are distilled by this method. In this process the temperature is gradually increased when the more volatile oils distill first and the less volatile ones distill progressively later. The distillate can be further purified by distilling it again and again.

2. **Expression.** The rinds of citrus fruits are subjected to great pressure to squeeze out the essential citrus oils. The oil mixed with the juices of the fruits is separated by centrifuging. Hand operated presses or crushers or large mechanical presses are used to express the essential oils.

3. **Fat Extraction.** Cold fat extraction or **enfleurage** is done at normal temperature. This method is generally employed in the case of flowers with delicate oils which are adversely affected by distillation, etc.

Pure tallow or lard which are odourless and are highly absorptive of floral essences are used to cover glass plates arranged one above the other in wooden frames. The perfumed flower parts are placed on these fat layers for few days, after which they are replaced by a fresh batch of flowers. After serveral dozen successive batches of flowers (throughout the blooming season) are allowed to saturate the fat with the essence, the fat layer *('**pomade**')* is subjected to alcoholic extraction. The essential oils being soluble in alcohol dissolve in it which is then concentrated to yield the **perfume oil**.

Sometimes fat extraction is done at high temperature when the process is called **maceration**. In this case small pieces of certain flowers are digested with hot oil or melted fat. The resulting pomades are subjected to alcoholic extraction.

4. **Solvent Extraction.** This is a comparatively recent process employed in the last one hundred years only (since 1879). In this process the perfumes are extracted directly by means of certain purified volatile solvents like petroleum ether, benzol or alcohol. The solvent is then allowed to evaporate away in vacuum leaving a semi-solid residue *(**concrete**)* of oils and waxes. The wax is removed by adding alcohol when the oils dissolve in it. The wax is then filtered off. The alcohol is also then removed and a concentrated form of perfume oil is obtained.

Classification of the Essential Oils

A water-tight classification of essential oil plants is not possible since many of them yield oils which have uses in perfumery, in flavouring, in medicine as well as in industries. However, the essential oils are classified into three general groups for the purpose of convenience.: (1) perfume oils, (2) flavouring (spice) oils, and (3) industrial, medicinal or special purpose oils.

The spice and other flavouring material yielding plants and the medicinal plants are discussed in other chapters.

PERFUMES

History. Perfumes have a very old history and find references in the records of the earliest civilizations. The Egyptians and the Hebrews are known to have used them for personal and religious purposes. Perfumes were used in the burial rites for Egyptian Pharoahs. The Chinese used the perfumes on their body and essences for flavouring their food. Fragrant incense was used by the Chinese, Indian, Persian, Jews and the Babylonians as offerings to their Gods. The Romans and the Greeks used perfumes at religious function. The rich Greeks used different perfumes for different parts of the body. At Versailles, King Louis XIV of France personally supervised the proper blending of essences for his royal bath, a particular essence formula for each day of the year. Later due to a very widespread demand for essence in France, perfume industry was created by royal charter. Today also perfume industry of France continues to be the leading one in the world followed by England, India, Turkey and United States.

The natural perfumes are usually adulterated with synthetic ones to make them slightly cheaper. The synthetic materials are, however, not very lasting. The costly perfumes are usually blends of several natural perfumes. The perfumes are normally mixed with less volatile oils, the fixatives, which may be of plant or animal origin (musk, ambergris, civet, etc.).

The perfumes are stored in closed, completely filled containers since they deteriorate due to oxidation and polymerization on coming in contact with air.

Examples of Perfume Oils

1. ***Rose oil*** (Otto, or Ottar, or Attar of roses). It is obtained from flowers of damask rose, *Rosa damascena,* grown chiefly in Bulgaria. They are also grown in France, Italy, north Africa, Asia Minor and India.
2. ***Geranium oil.*** It is obtained from the leaves of several species of *Pelargonium* of the family **Geraniaceae**. They are chiefly grown in southern Europe and northern Africa.
3. ***Neroli oil*** or ***Orange blossom oil.*** It is obtained from the flowers of certain species of orange, particularly from bitter orange *(Citrus aurantium)* of **Rutaceae**.
4. ***Yiang-yiang oil*** (cananga oil). It is obtained from the petals of the flowers of *Cananga odorata* of the family **Annonaceae**. It is grown in the Philippines, Vietnam, Cambodia and other parts of the southeastern Asia.
5. ***Cassie oil.*** It is obtained from the flowers *of Acacia farnesiana of* **Leguminosae**. It is chiefly grown in France, Algeria, Egypt, Syria and India.
6. ***Lemon-grass oil.*** It is obtained from the leaves of *Cymbopogon citratus* of the grass family, **Poaceae**. It is grown in India, Sri Lanka, Congo, Madagascar, Guatemala, Haiti, Brazil, Salvador, etc.
7. ***Lavender.*** It is obtained from the flowers of *Lavandula offlcinalis* of **Lamiaceae**. It is grown in France and England.
8. ***Jasmine***. The essence is obtained from the flowers of *Jasminum officinarum* var. *grandiflorum.* The plant is grown in southern France and other Mediterranean countries.
9. ***Champaca oil.*** It is obtained from the flowers *of Michelia champaca.* It is a famous perfume of India.
10. ***Sandalwood oil.*** It is obtained from *Santalum album* and allied species.
11. ***Saffron oil.*** It is obtained from the dried stigmas of *Crocus sativus* of **Iridaceae**. It is grown in the Mediterranean area and the Asiatic tropics.

12. ***Narcissus oil.*** It is obtained from *Narcissus jonquilla, N. poeticus* of **Amaryllidaceae**. It is cultivated in southern France.
13. ***Khus-khus oil.*** (Vetiver oil). It is obtained from the roots of a grass, *Vetiveria zizanioides*. It occurs wild in many parts of India. It is cultivated in Java, Reunion, the West Indies and in the southern United States of America.

MEDICINAL AND INDUSTRIAL OILS

Camphor and Camphor Oil

Camphor oil is liquid at higher temperatures, but on cooling the substance called camphor separates out. Camphor is a colourless, tough crystalline solid at ordinary temperature with a sharp odour and pungent, aromatic taste. It volatises very slowly. It is used in the production of celluloid, and various nitrocellulose compounds and is also of medicinal value. In India it is widely used in religious functions. It is burnt before the deities of Gods.

Camphor is derived chiefly from the stems, twigs and leaves of the camphor tree, ***Cinnamomum camphora*** of **Lauraceae**. The plants is native of **China**, **Formosa** and **Japan**. The cultivation of the tree is now restricted to Formosa and Japan due to production of synthetic camphor in U.S.A. and England. The tree can grow well at elevations of up to 3,000 to 4,000 feet and between 44°N and 45°S latitudes.

The plant is a large tree which can attain a height of a hundred feet or more. It is, however, of smaller height when cultivated. The camphor is deposited as a solid in the cells.

Plants which are 50 or more years old are used for the extraction of camphor. The wood, twigs and leaves are macerated and distilled with steam. The crude crystalline camphor which separates out from the camphor oil on cooling is purified further by sublimation with quick lime and charcoal. The residual oil contains an artificial sassafras oil, **safrole**, which is a more valuable product than camphor. It is used for the production of **heliotropin**.

Cedarwood Oil

It is obtained from the wood of eastern red cedar ***(Juniperus virginiana)*** of **Cupressaceae**. The plant is native to eastern **United States**. It is a cheap oil. It is used as a cleaning agent in preparing permanent slides. It is used with oil-immersion lenses, in perfumery, soaps, deodorants and as an adulterant of expensive essential oils.

Peppermint Oil

It is obtained from some species of ***Mentha (M. piperata)*** of **Lamiaceae**. It is indigenous to **Europe**. It is now cultivated in European countries, U.S.A., Japan and India and many other countries. The oil has a content of methanol.

It is used for flavouring chewing gums, sweets, tooth pastes, etc. It is of great medicinal value and is used as an antiseptic lotions and in digestive disorder.

Ajowan Oil

It is obtained from the seeds of ***Trachyspermum copticum*** of **Umbelliferae**. The plant is a small annual herb of India. It also grows in Egypt and Persia. The oil has a high thymol content. It is of medicinal value and is used in digestive disorders. The fruits are used as constituents of curry powders.

QUESTIONS

1. What are essential oils. Why are they called essentials? Name five essential oil yielding plants along with their botanical name.

20

Medicinal Plants

HISTORY

Medicinal plants were known to the early civilizations. As a matter of fact the history of the drug plants is as old as the history of these civilizations. The Chinese are reported to have used drug plants as early as 5,000 to 4,000 B.C. The Assyrians, Babylonians, Hebrews and Egyptians knew many drug plants in about 1600 B.C. The works of Greeks, *viz.,* **Aristotle** (384–322 B.C.), **Hippocrates** (460–370 B.C.), **Pythagoras** and **Theophrastus** (370–287 B.C.) have numerous references of many of the present day drugs. In 77 B.C., a Roman physician **Dioscorides** wrote **'De Materia Medica'** which described the nature and properties of all the 500 medicinal plants known at that time. His book was considered to be the most authentic work on medicinal plants for the next 16 centuries or so. There was no advancement in the knowledge of drug plants during the Dark Ages. After the introduction of printing in Europe in the fifteenth century many persons published '**herbals**' which contained many true and false informations. Some people advanced the 'Doctrine of Signs or Signatures' which meant that the appearance of a plant or its organs indicated its utility, the sign being placed there by the Creator. Mandrake ***(Mandragora officinarum)*** was one such plant, the human body like appearance of which led to the superstition that it was useful in the treatment of human diseases.

In the present times medicinal science has paid great attention to the study of drug plants. The branch of medical science which deals with the drug plants is called **pharmacognosy**, whereas the study of the action of drugs is called **pharmacology**.

The drugs are usually classified on the basis of the morphology of the drug-yielding parts of plants.

DRUGS OBTAINED FROM ROOTS

ACONITE

The drug is obtained from the tuberous roots of ***Aconitum napellus*** of **Ranunculaceae**. The plant is a perennial herb. The plant is also known as **monk's hood** or **wolfbane** (Fig. 20.1). It is indigenous to the **mountains of Europe** (Alps, Pyrenees, etc. and western Asia. It is also cultivated as an ornamental plant in most of the countries.

Fig. 20.1. ***Aconitum napellus***. A portion of the plant.

Several alkaloids are obtained from the roots. The most important alkaloid is **aconitine** which is poisonous. Aconite is usually used externally for the treatment of **neuralgia** and **rheumatism**. Internally it is taken to relieve pain and fever.

RAUWOLFIA

Origin and History

The drug Rauwolfia is obtained from the roots of ***Rauwolfia serpentina*** of **Apocynaceae**. It is native of **India**. The genus is found growing in the tropical regions of Asia, Africa and America. The Asian countries where ***Rauwolfia*** grows in abundance are India, Bangla Desh, Sri Lanka, Burma, Malaysia, Thailand and Indonesia. Five species of ***Rauwolfia*** grow in India. ***R. Serpentina,*** the most important of the five species is found in the sub-Himalayan tract from Punjab through Nepal, Sikkim, Bhutan to Assam, eastern and western ghats, Central India and the Andamans. The plant is grown commercially in Uttar Pradesh, Bihar, Orissa, West Bengal, Assam, Andhra Pradesh, Tamil Nadu, Karnataka, Kerala and Maharashtra.

Fig. 20.2. Inflorescence of *Rauwolfia.*

Botanical Description

The plant is an erect perennial shrub. It has a height of $\frac{1}{2}$ ft. to $1\frac{1}{2}$ ft. and may sometimes attain a height of 3 ft. The plant is evergreen. Leaves are in whorls of three. Large number of small white or pink flowers are born in cymes (Fig. 20.2). The fruit is a drupe. It is propagated by seeds.

Ecological Factors

The plant thrives well in hot and humid climates of the **tropics**. It prefers shady habitats of the forests. It grows well in areas of very high rainfall (250 cm. to 500 cm.) and temperature range of 10 to 38°C. It can grow in different kinds of soils ranging from sandy alluvial loam to red lateritic loam. Well drained humus rich clayey soil is best for its growth.

Harvesting and Extraction

The drug is extracted from the barks of the roots, which are 2 to 3 years old. The roots are preferably harvested after leaf-fall because of their higher alkaloid content (90%). The stems and leaves have negligible amount. The roots are cleaned and dried before the actual extraction process.

In India Rauwolfia has been successfully used for extracting a number of alkaloids such as ajmaline, ajmalinine, ajmalicine, serpentine and surpentinine from ***Rauwolfia serpentina***.

Medicinal Properties

The roots of the plant constitute the rauwolfia drug. The bark of the roots yield several alkaloids (**about 80**), the most important of which is **reserpine**. Reserpine is extracted from another two species—***R. vomitoria*** of Africa and ***R. tetraphylla*** of America. Other important alkaloids are **deserpidine**, **rescinnamine**, **reserpinine**, **serpentine**, **serpentinine**, **ajmaline**, etc.

Uses of Rauwolfa

1. Reserpine is extensively used in the treatment of violent kinds of insanity and high blood pressure. It acts as a sedative and depressant in hypertension, epilepsy and chronic psychoses.
2. It is also used for treatment of insect bites, fevers and dysentery. Reserpine is now widely used by research workers to study physiological systems.
3. It is used for inducing uterine contraction at the time of childbirth.
4. The juice extracted from leaves is useful in removing the opacity of the cornea.

DRUGS OBTAINED FROM BARKS

QUININE

Origin and History

Quinine is obtained from the bark of several species of ***Cinchona*** of **Rubiaceae**. The plant is native to **Andean highlands of tropical America**. It still grows in Peru and Bolivia at altitudes of 3,000 feet to 9,000 feet.

The medicinal properties of the bark of the tree were discovered in 1638 when it cured the malarial fever of the wife of the Viceroy of Peru, the Countess of Cinchon. She carried it to Spain in 1639. The plant was named ***Cinchona*** by Linnaeus in the eighteenth century after her name. The bark of the plant called Jesuit's bark or Peruvian bark attracted the attention of the people throughout the world. From mid-seventeenth to the mid-nineteenth century the wild trees in

South America were ruthlessly cut down to get the bark. The British and the Dutch secured the seeds from South America and started large plantations in India and Java respectively. The Britishers had collected seeds from ***Cinchona succirubra***. The Dutch plantations of Java were established from the seeds of ***Cinchona ledgeriana*** named after a British resident of Bolivia, Charles Ledger, who had collected the seeds. Java accounts for 90 per cent of the world export trade in quinine. Other species of ***Cinchona*** used for the drug are ***C. officinalis*** and ***C. calisaya***. The plant was also grown in Sri Lanka but was replaced by tea plantations due to uneconomic price of the drug. The other *Cinchona*-growing regions are Burma and Tanganyika (Tanzania).

Botanical Description

Cinchona plant is a fairly large tree and may attain a height of fifty feet or more. The stem is covered with thick bark which may be light coloured, pale, brown or dark brown depending on the species. The leaves are opposite. Yellow or pink coloured flowers are borne in terminal panicles. They are produced from the third or fourth year of growth. Cross pollination takes place.

Ecological Factors

The plants are limited in distribution. They grow in the tropics between 10° north and 20° south of the Equator at altitudes more than 1,000 feet. They grow well in regions of very high temperature and high rainfall (60 inches or more). Good porous soils are good for their growth.

Cinchona is propagated by seeds. Natural hybridization has resulted in mixing of characters and consequent variation in plants of different plantations. It is now usually propagated by means of graftings and cuttings.

The trees are usually felled down after they are 10 years old since the bark develops maximum alkaloid content by that time. The bark is removed from the stems, branches and roots and is slowly dried at not very high temperature. The dried bark is sent to the factory where the alkaloids are extracted by solvent extraction.

Medicinal Properties

Cinchona bark has four important alkaloids (**totaquine**) which are **quinine**, **quinidine**, **cinchonine** and **cinchonidine**. Quinine is the most important of the alkaloides. ***C. ledgeriana*** has the maximum percentage of quinine in relation to the other alkaloids.

Quinine is a very bitter, white granular antimalarial drug. It is also useful as a tonic and antiseptic. Some synthetic antimalarial drugs are also manufactured these days.

DRUGS OBTAINED FROM STEMS

EPHEDRINE

The drug ephedrine is an alkaloid. It is obtained from several species of ***Ephedra*** of **Ephedraceae** of **Gnetales**. Two important species are ***E. sinica*** and ***E. equisetina***. The plant is native of **Asia**. In China it has been in use for over 5,000 years. ***Ephedra*** is world wide in distribution but is chiefly grown in India, China, Spain, U.S.A, Afghanistan, Tibet, Australia, and Kenya. In India the plant grows in Western Himalayas and Sikkim.

The plant is leafless shrub with green stems. The plant is woody and xerophytic and grows in arid regions of the world. The plant is dioecious.

The alkaloid is extracted from the entire plant. Synthetic manufacture of the drug has reduced the importance of ***Ephedra*** as a source of ephedrine.

Ephedrine is extracted from fresh or even dried branches of the stem, when the plant is nearly four years old.

Ephedrine is the chief alkaloid present in the plant (75%). The other alkaloids occurring in ***Ephedra*** are pseudo-ephedrine and ephedine.

Uses of Ephedrine

1. Ephedrine and its synthetic derivatives are widely used for treating asthma, hay fever and cold.
2. Ephedrine is used for treating nasal, sinus and bronchial congestion.
3. Ephedrine is also used as a stimulant of the heart.
4. It is used for treating rheumatism and syphilis.
5. The root extract is used a tonic and diuretic.
6. It is used to treat bed–wetting by children.

DRUGS OBTAINED FROM LEAVES

BELLADONA

Origin and Historhy

The drug is obtained mainly from the dried leaves of the deadly nightshade, ***Atropa belladona*** of **Solanaceae**.

The plant is indigenous to **Central** and **Southern Europe** and **Asia Minor**. The plant is widely cultivated in India, Europe and U.S.A. In India it is chiefly grown in Kashmir and Uttarakhand.

Botanical Description

The plant is a perennial herb with a creeping root stock. It bears alternately arranged ovate leaves. The stem is hollow. The flowers are purplish. The plant bears brownish or blackish berry. The seeds are sown in warm and shaded habitat. The plants grow best in calcareous, dry loamy soil rich in humus, sodium and potassium.

The flowering season is the best for harvesting, since the plant has highest concentration of alkaloids at that time. The tops and leaves of the plants are dried for sometime. The alkaloids are extracted mainly from the leaves of which **atropine** and **hyoscyamine** are more important. They have been found to be useful in acting as antidote to opium, in reducing nasal secretions etc. The leaves are then dried at low temperatures for a period of 2 to 15 weeks. The alkaloids present in the plant are extracted with the help of solvents.

Medicinal Properties

The plant yields several alkaloids-**atropine** ($C_{17}H_{22}O_3N$) **hyoscyamine**, **scopolamine**, **apoatropine**, **belladonine**, **norhyocyamine noratropine**, **hyoscine**, **tropacocaine** and **meteloidine** out of which the first three are of greater importance. They are used as a stimulant to the sympathetic nervous system, as diuretics, in **dilating the pupil of the eye** and in the treatment of palsy. Externally it is used in ointment to relieve **pain**. Atropine or hyoscyamine is used to dilate the eye, check excessive perspiration, stimulate circulation, relieve local pain and to counteract muscle spasms. **Scopolamine** is used as a **narcotic**, **anti-insomniac** and **anesthesia**.

Atropine is a stimulant of respiratory and circulatory system. **Scopolamine** reduces the parasympathetic nervous system. It induces sleep since it acts as a sedative. Belladona is useful in controlling whooping cough and asthma. Externally it is used for the treatment of gout and rheumatic inflammation.

The roots and leaves are used as narcotic and diuretic. The berries are very poisonous.

COCAINE

Origin and History

Cocaine is obtained from the leaves of a **South American plant**, ***Erythroxylon coca*** of **Erythroxyllaceae**. The plant is indigenous to the **Andean regions of Peru** and **Bolivia**. The plant is now widely cultivated in the tropical regions of South America, Java, Sri Lanka and Taiwan (Formosa). The plant grows at higher altitudes.

Botanical Description

The plant is a small, much branched shrub or tree. It bears small, sessile, elliptical leaves which are alternately arranged. The flowers are borne is small axillary groups (Fig. 20.3).

Fig. 20.3. ***Erythroxylon Sp.*** A twig of coca plant.

The leaves are harvested by hand when the trees are two to three years old. The leaves are picked three or four times in a year. Dried leaves are immediately exported in boxes.

Medicinal Properties

The alkaloid cocaine ($C_{17}H_{21}O_4N$) is usually extracted in the importing countries by **solvent process**. The drug is chiefly used as a local **anesthetic**. It is also used as a tonic for the digestive and nervous symptoms. The leaves are chewed by the natives of South America for stimulating physical and mental activities. The coca is 'de-alkaloidized' in U.S.A. to be used as 'cola' flavourings.

DIGITALIS

Origin and History

The two common specis are ***D. purpurea*** and ***D. lantana***. They are widely grown Australia, Germany, Hungary and also in U.S.A., England, Czechoslovakia and Argentina in South America. In India the plant is grown in Kashmir and Nilgiri hills, Uttarakhand and hills of Darjeeling on a limited scale.

The plant is a native of **Southern and Central Europe**, which continues to be its chief grower. The plant is also grown in other parts of the world as beautiful ornamental plants.

Digitalis pururea (foxglove) belongo to the family **Scrophulariaceae.** The plant is a pubescent biennial or perennial herb. The leaves are crowded in the lower portion of the stem which is about 5 feet in height. The stem bears purplish flowers in a terminal spike. The leaves are dried in shaded conditions. The active **glycosides** are extracted from the dried leaves by **solvent (alcohol) process**. The glycosides are **digitoxin**, **digitaline** and **digitalein** and **digiton**. They are very important as stimulants of heart and as diuretics. Digitalis induces the **heart** to make powerful and complete contractions so that the circulatory system is toned up.

The chief source of the drug, the large sized leaves of the plant, must be harvested in the first year before the flowering season. The leaves can be dried either by artificial heating or by spreading them in sunlight.

The glycosides have been found to occur within the collenchymatons tissue of the epidermal region, as well as in the endodermis. The cardiac glycosides are lanatosides and saponins.

Digitoxin has been found to be 1000 times more effective than powdered digitalis. The derivatives of the primary glycosides i.e. lanatosides A, B and C are digitoxin, and digoxin. If further hydrolyzed.

Eucalyptus

Eucalyptus oil is used in medicine. It is obtained from the leaves of several species of *Eucalyptus*. The plant is native to **Australia**. It is also grown in the Mediterranean region, U.S.A. and many other places.

The plant is a tall tree which attains a height of 200 to 300 feet. The leaves are scythe-shaped in *E. globulus* and of different shapes in other species.

Eucalyptus oil is used in the treatment of cold, malaria, nose and throat troubles and fevers.

Fig. 20.4. Eucalyptus Tree.

DRUGS OBTAINED FROM FLOWERS, FRUITS AND SEEDS

STRYCHNINE

The drug is obtained from the seeds of ***Strychnos nuxvomica*** of **Loganiaceae**.

The plant is indigenous to **Australia**, **India**, **Sri Lanka** and **China**. The plant also grows widely in South America and Europe. It was introduced is Europe in the 16th century. The plant is a woody vine of the tropical forests. The stem bears ovate leaves in opposite manner. The plant bears small flowers and large indehiscent fruits. A fruit of ***Strychnos*** contains 3 to 5 hard and greyish seeds. The alkaloids present in the seeds are **strychnine** ($C_{21}H_{22}N_2O_2$) and

brucine ($C_{23}H_{26}N_2O_4$). They are expressed from the seed with boiling sulphuric acid. The alkaloids are first precipitated and are then purified.

Strychnine is a very poisonous substance. It is used medicinally in the treatment of **nervous disorders** and **paralysis**. It stimulates the central nervous system and acts as a tonic. It is famous as **arrow poison (curare)**.

OPIUM

Origin and History

Opium is obtained from immature capsules of opium poppy ***(Papaver somniferum)*** of **Papaveraceae**. The plant is indigenous to **Asia Minor** and **India**. It is also grown in China, West Asia and the Mediterranean region.

The opium poppy plants are widely cultivated in India, Turkey, Afghanistan, Pakistan, Russia, and Japan. In India certain parts of Madhya Pradesh, Uttar Pradesh and Rajasthan are involved in their cultivation. It is a winter crop and grows best in well drained sandy loam.

The plant is also grown as an ornamental one.

Botanical Description

The plant is an annual herb of a height of 2 to 4 feet. The plant bears alternate leaves and white, pink or reddish-purple flowers. The fruit is a globular pale green capsule (Fig. 20.5).

Fig. 20.5. Photograph of opium plants with their fruits.

The gynoecium is tricarpellary, syncarpous, one-chambered and has parietal placentation. The buds are always drooping but the flowers have an upright stalk.

Poppy capsules have pores in their neck for dispersal of seeds. The dehiscence is of porous type.

Extraction

A number of fine parallel slits are made into the unripe capsules in the evening and in dry weather. The exuding **latex** is collected in the morning. It is rolled into balls which are covered by dried petals. Whereas the '**alba**' varieties are grown in India and China for getting commercial opium, the '**glabra**' varieties are used for extraction of the alkaloids to be used in medicines.

Medicinal Properties

There are about **30 alkaloids** out of which the important ones are **morphine**, **codeine**, **narcotine** and **papaverine**.

Morphine and codeine are widely used as **sedatives** to relieve pain and cause sleep. They are usually taken orally, rectally or by injections to cause local insensitiveness. Morphine is also used in the treatment of cough.

Opium is a very valuable as well as a dangerous drug. Opium has narcotic effects. It is eaten and smoked by millions of people for deriving pleasant feelings of exhilaration and peace. Addiction to opium, morphine, codeine, **heroine** (artificial derivative of morphine) etc., however, leads to physical and mental degradation, debility, delirium and even death.

The drug should be used in very limited quantities and under strict supervision of a physician.

Uses of Opium

1. The heroine is a more powerful analgesic than morphine.
2. Morphine is component of several cough medicines.
3. It controls diarrhoea and vomiting.
4. Codeine is well known pain reliever. It is given to relieve whooping cough.
5. The seeds of opium poppy are consumed in many ways. They are mostly used for cooking stuffed "parathas".
6. The seeds are also used for extracting poppy oil.
7. Morphine is the most powerful of all the alkaloids of opium (8 – 13%). While codeine is a milder analgesic, heroin (a derivative of morphine) is a much more powerful analgesic than morphine. Morphine is useful in controlling cough, diarrhoea, vomiting, bleeding and blood pressure.
8. Poppy seeds do not have any harmful effect. The seeds are edible and are relished by Indians either in the form of stuffed breads or in cakes etc.

DRUGS OBTAINED FROM THE LOWER PLANTS

ANTIBIOTICS

Antibiotics are produced by molds, actinomycetes and bacteria. They kill pathogenic bacteria. Some of the antibiotics produced by such microorganisms are as follows:

Penicillin is obtained from *Pencillium notatium* and *P. crysogenum.*

Streptomycin is obtained from *Streptomyces griseus.*

Aureomycin is produced by *Streptomyces aureofaciens.*

Chloromycetin is obtained from *Streptomyees venezuelae.*

Terramycin is obtained from *Streptomyces rimosus.*

Neomycin is obtained from *Streptomyces fradiae.*

OTHER DRUGS

Agar is obtained from red algae, *e.g., Gelidium corncum, Eucheuma spinosum, Gracilaria lichenoides* and some other species.

Ergot is obtained from the fruiting body of a fungus, *Claviceps purpurea.*

Kelp comprise certain brown algae, *viz., Macrocystis pyrifera, Laminaria digitata, L. saccharina,* etc. They yield iodine, potash, acetone, sodium alginate, etc.

INSECTICIDE YIELDING PLANTS

In view of the fact that chemical insecticides like D.D.T. are also toxic to the human beings, search for non-toxic botanical insecticides has been going on all the time. Over 1200 species of plants have been found to be insecticidal. **Pyrethrum** and **rotenon** are two important insecticides obtained from plants.

Pyrethrum

Pyrethrum is obtained from the flowers of three species of ***Chrysanthemum*** of which the insect flower plant ***C. cinerariaefolium*** is the most important. They belong to the family **Asteraceae**. The plant is native to southeastern Europe. The plant is cultivated throughout the world. Japan and British East Africa are the chief growers of the plant. The other important growing regions are California and some other parts of the United States, Italy, Australia, Brazil, Peru and Equador.

Chrysanthemum is a small, perennial herb attaining a height of 2-3 ft. It belongs to the family **Asteraceae**. The inflorescence is typically head or capitulum, which has hermaphrodite disc florets and female ray florets. The plant requires well-drained soil. The soil may be sandy or clayey: The plant grows best in mountaineous regions where low temperature stimulates flowering.

The Insecticide

Pyrethrins are synthesised at their optimum concentration just before the complete opening of the florets. The pyrethrins are of two types—pyrethrin I and pyrethrin II. The ovaries contain 90% of the insecticide. The two pyrethrins were discovered by **Staudinger** and **Ruzicka** (1924). Pyrethrins also include **cinerin I** and **cinerin II** which were discovered by **La Forge** and **Haller** (1947). The insecticide is highly effective against pests of various types and its mode of action is extremely rapid. It is neither toxic to animals nor is inflammable and is, therefore, in great demand.

The insecticide is extracted from the flowers with the help of kerosene and is used as a spray. It can also be extracted from the flower by using ethylene dichloride as a solvent. Yet another method for extracting the insecticide is by dissolving the residues in kerosene and other solvent when the preparation can be used as a spray. The pyrethrins are also extracted in the form of a solution which is mixed with substances like sesamin etc to be used as aerosol sprays. If mixed with some herbicides the insecticide becomes much more effective.

Pyrethrum is widely used as spray against house-flies, moths, mosquitoes, body lice, cockroaches etc.

The insecticide is not poisonous. It is also non-inflammable. It kills fleas, flies, mosquitoes, lice, etc. Pyrethrum ointment is used in the treatment of scabies.

ROTENON

(Lonchocarpus utilis** and **Derris elliptica)

Origin and History

Rotenon yielding plant ***Derris*** is native of **Malaya**, and **South East Asia**. They are now cultivated in Sumatra, Java, North Borneo, the Philippines, Tanzania, Zaire, South America, and West Indies. In India it is cultivated in Assam, Karnataka, Kerala, Tamil Nadu and Punjab.

Lonchocarpus is believed to be native of **Brazil**, **Peru**, **Guyana**.

Botanical Description

Derris belongs to the family **Fabaceae**. It is a woody climber. ***Lonchocarpus*** which is initially bushy becomes climbers. They grow best in tropical climate where rainfall is high and uniform almost throughout the year. They require well-drained soil. They are propagated by means of cuttings.

Harvesting

The insecticide is obtained from the roots of the species of these two plants. Two to three year old roots are taken out of the ground and are kept in the open so that they dry up. Bundles of dried up roots are processed to extract rotenon either in the form of powdery substance or in the form of liquid.

The Insecticide

Rotenon is not soluble in water but is soluble in organic solvents like acetone, alcohol, chloroform, ether, benzene, etc. Rotenon is very prisonous but is harmles for animals and haman beings.

The chemicals obtained from the roots of ***Derris*** and ***Lonchocarpus*** are rotenon, elliptone, deguelin, toxicarol, sumatrol, and malaccol. The chemicals are generally present within the vascular rays, the cells of the pericycle and parenchyma.

QUESTIONS

1. Give botanical names of plants yielding drugs for hypertensions, asthma and heart ailments. Discuss the morphology of any one of them.
2. How was quinine discovered? Explain.
3. Name the alkaloids obtained from *Rauwolfia serpentina* and *Papaver somniferum*.
4. Give a brief account of any two of the drug plants you have studied with special references to their uses in pharmaceutical industry.

21

Fumitories and Masticatories

(Smoking and Chewing Materials)

Man has been using smoking and chewing materials since times immemorial for the pursuit of pleasure. Most of these materials have a stimulating or narcotic effect due to the presence of alkaloids. The narcotic plant products and their derivatives are extensively used the world over. Their continued use make people drug addicts. These materials have very harmful effects on them. The addicts lose their senses temporarily and sometimes even permanently. Such people are problems for family as well as for the society.

TOBACCO

(Nicotiana tabacum)

Tobacco is extensively used throughout the world for smoking and chewing purposes. Dried, cured and fermented leaves are used because of the stimulating and slightly narcotic effects.

Origin and History

Tobacco is indigenous to **tropical America**. The plant has been under cultivation from ancient times. Its original wild ancestor is not known. The plant was introduced into North America long before Columbus discovered it. He saw the natives of Cuba using tobacco in 1492. Spanish explorers introduced it into Spain in 1558. It was carried to England in 1585 by **Sir Walter Raleigh**. By the end of the sixteenth century and early seventeeth century the plant spread to other parts of Europe, Africa, India, China, Sri Lanka and Australia. Tobacco is now cultivated in U.S.A, India, China, Rhodesia and parts of East Africa, the Balkan States, Russia, Japan and Brazil.

Botanical Description

Tobacco plant belongs to the family **Solanaceae**. It is a small herbaceous annual. In the early stages it appears as a rosette but grows to a height of three to five feet. It has a well developed and strong tap root system. The stem bears very large ovate leaves. The inflorescence is a terminal raceme of white or pink flowers. The capsules contain numerous small seeds.

Commercially two species of *Nicotiana* are important–*N. tabacum* and *N. rustica* of which the former is by far the more important one. *N. rustica* occurs wild in Mexico and in cultivated to some extent in parts of Europe, Northern Asia and Indonesia. The leaves of this plant have a high nicotine content and the plant is mainly grown as a source if drug and for insecticidal purposes.

Ecological Factors

Tobacco is a tropical crop but grows well in sub-tropics and even in temperate climates. Moderate rainfall evenly distributed throughout the year is good for the crop.

The crop thrives best in a well-drained light sandy loam. The soil with rich humus and potash, lime and other essential elements is good for the plants.

Cultivation

Tobacco seeds are sown in seed bed by broadcast method. The seedlings are transplanted in well cultivated fields. When the plant produces eight to twelve leaves, the top of the plant is removed ('**topping**') as a result of which the axillary shoots are developed. These shoots are then removed to encourage increase in the size, weight, thickness and strength of the leaves.

The leaves of the plant mature in four to five months time. The leaves turn greenish yellow during ripening. The leaves are harvested as they ripen from below upwards for giving best quality tobacco. Very often the whole plant is harvested. Different methods of curing determines the time of harvesting with different degree of ripeness of the leaves.

The leaves of tobacco contain about 70 to 80 per cent water, **starch**, **nitrogenous compounds**, **organic acids** and 1 to 8 per cent of an alkaloid, **nicotine**.

Curing of Tobacco

The smell and taste of tobacco develop during the curing process due to the liberation of various **essential oils** and other **aromatic substances**.

The harvested plants or leaves are stacked or hung in curing barns. Curing is of four kinds. **Air-curing** is done in well ventilated barns under moist and warm conditions. It is a slow fermentation process. **Flue-curing** is a faster process. It involves heating of the barns with the help of furnaces. **Fire-curing** is done by smoking from slow fires of charcoal etc. without any appreciable rise of the temperature. **Sun-curing** is performed in the open in the Asian countries.

Curing involves dry fermentation of the leaves which lose water, green colour, starch and sugars. Protein (about 60%) is lost in the form of **ammonia gas**. The amount of oxalic acid is not changed but citric acid increases. The volatile form of nicotine evaporates and, therefore, the amount of nicotine (only the non-volatile form remains) decreases. Curing takes three to six months time.

The cured leaves are then sorted and similar grade leaves are tied into 'hands'. The graded leaves are placed in heaps or are pressed into special containers for the purpose of fermentation. The fermentation is done in large warehouses. The process is completed in six months to four years time. During this process further changes occur in the chemical composition of the leaves. The nicotine content is further decreased. The characteristic odour and flavour of the tobacco develop during this period. The harshness, bitterness and other undesirable qualities disappear and the burning qualities of the tobacco is enhanced.

Tobacco is used for chewing, snuff and smoking purposes. Tobacco is smoked in a number of way. While the villagers use earthen pots called '**Chilam**' and '**Biri**' for smoking tobacco, the urban people use cigaretters, cigars, pipes etc.

BETEL AND ARECA NUT

(Piper betle* and *Areca catechu)

Origin and History

Betel or **pan** is a very popular chewing material in the Asian countries. It is used extensively by the people of Zanzibar, India, Burma Malaysia, Indonesia, Vietnam, Thailand, Cambodia, South-China and the Philipines from the remote past. An average Indian particularly in the North is very fond of betel and consumes it several times in a day.

Botanical Description

The betel pepper is a perennial, creeping climbing plant. It is a native of **Malaya** and is cultivated throughout the southeastern Asia. The dark green leaves have a hot astringent taste. They are broad with pointed tips. They are borne alternately on the stem. The plant is grown under shaded conditions to cause an increase in the size of the leaves.

The Areca palm is a very tall slender palm. It is grown throughout Asia for its seeds. The tree has crown of large pinnate leaves at the top. The inflorescences consist of much-branched spadix. The fruit is an ovoid drupe. The ripe fruit is orange coloured. The pericarp is a hard fibrous structure. It encloses a single seed. The seeds contain an alkaloid **arecoline**. Fresh or cured seeds are used after slicing them into small pieces. Cured seeds are better than fresh seeds because they have lower tannin content, better palatability and better storage properties.

Betel has four principal ingredients-leaf of betel ***(Piper betle)*** seeds of Areca palm ***(Areca catechu)***, lime and a resin called 'katha'. A mixture of areca nuts, 'katha' and slaked lime is rolled in a leaf of betel vine to prepare pan. Betel is often chewed along with some other aromatic substances as well. These are cloves, fennel cardamom, etc. Tobacco is also quite often used as an ingredient of betel.

Betel is chewed slowly in the mouth. It causes flow of saliva which is spit out every few minutes. It has stimulating and mildly narcotic effects on the human body. It is also believed to have some medicinal value.

COLA

The cola or kola nuts are the seeds of the cola tree ***(Cola nitida)*** of **Sterculiaceae**. They are widely used as a chewing material as well as a source of flavour and caffein in **'cola' drinks**. The cola tree is indigenous to **tropical western Africa** but is now grown in Sudan, Jamaica, Brazil, India, etc. It is a forest tree. The fruits are star-shaped. A fruit contains 8 reddish seeds.

The chewing of fresh cola nuts has stimulating effect, and inhibits fatigue and hunger. It contains caffein, an essential oil, and a glucoside—kolanin. Caffein and kolanin are responsible for the stimulating effect of the cola seeds.

THE TRUE NARCOTICS

There are some plants which contain alkaloids of great economic importance. In very small amounts these alkaloids are used in medicines for relieving pain, anxiety, tension, etc. and to induce sleep. Cocain and opium are examples of such narcotics. Hashish, peyote, cappi, etc. are, however, very strong narcotics. They have hypnotic and halucinating effect on the mind of the user. There is so much excitation of the brain that the person sees unreal visions and illusions. The drug habit results in great physical and mental degradation, senselessness, coma, convulsions and absurdity.

COCA

The leaves of coca plant *(***Erythroxylon coca***)* are used for obtaining the drug ***cocain*** and also for chewing purposes.

The plant is a shrub. It is native to **Peru** and **Bolivia**. It is now grown in South America, Java, India, Sri Lanka and Formosa.

The leaves mixed with lime and alkaline ashes of some plants are chewed to stimulate the physical and mental activities. The hunger and pain etc. are forgotten under its influence. Excessive chewing results in loss of health, disease and disability.

OPIUM

Opium has been dealt with as medicinal plant in the previous chapter. Opium smoking and eating has been prevalent in the Asian countries for centuries. It is now used in large number of countries. It is used for its strong narcotic effects. It is a very dangerous drug. It consists of alkaloids morphine, codeine, narcotine and papaverine.

Opium is the dried juice which exudes from the immature capsules of the opium poppy ***(Papaver somniferum)*** of **Papaveraceae**. The plant is native of **Asia Minor** or **India**. It is a herbaceous annual and flowers in three months' time only.

While Indians eat opium, the Chinese smoke it. The inhaling of fumes has worse effects on the human system.

The alkaloids, morphine and codein are extensively used in medicines for the relief of pain. The seeds of poppy are edible. They contain 4 to 50 per cent of a drying oil which is used in India for the preparation of sweetmeats and in curry. The oil is also used in the manufacture of paints and soap.

GANJA, BHANG, MRIJUANA, CHARAS, HASHISH

(Cannabis sativa)

As stated earlier in Chapter 10, the hemp plant ***(Cannabis sativa)*** *has* three kinds of varieties under cultivation—for **fibers**, for a **drying oil** and for a **narcotic drug**. The drug yielding variety is extensively cultivated in India, Nepal and the West Asian countries for thousands of years. The Chinese are reported to have used the drug in 3000 B.C.

The narcotic stimulant or the drug or the resinous oil is obtained from the **resinous hairs** of the inflorescence of the **female plant**. It is used for relieving pain and for treating nervous disorders. The drug contains powerful alkaloids.

The drug is prepared in different ways in different parts of the world. It is smoked as well as consumed as a beverage. **Ganja** is produced from the **dried** and **powdered** young female **inflorescences** and young **leaves**. It is usually smoked. **Bhang** or **Marijuana** (American name) is consumed along with water, milk or sweetmeats. It consists of tops of wild plants. The highest concentration of the drug is obtained from the resinous secretion occurring from the female inflorescences before fertilisation takes place. This pure resin is called **charas** or **hashish** (West Asian name). The active principle is **tetrahydrocannabinol**.

The various forms of the hemp listed above are all extremely dangerous narcotics for the human beings. They produce very deleterious psychological effects. It causes great amount of cerebral excitement, hypnotic effect, hallucinations, erotic visions, imaginary flights into the world of unreality, mental unconsciouseness, stupefaction, ecstasy and the like. The modern society of the western countries is today plagued with such addicts.

General Questions

1. Write an essay on the role of plants in everyday life, drawing your examples from as many sources as possible.
2. Name *any five plant* species which yield economically useful fibres. Write a note on the use of these fibres and indicate the morphological nature of the fibre in any *two.*
3. Suppose that all plants of the world were to be destroyed except those belonging to five plant famlies. Which famlies would you select and why?
4. (*a*) Mention the botanical name and economic uses of jute. Where does it grow in India?

   (*b*) Give the botanical name of any fibre-yielding fruit and indicate the use of the fibre.
5. Name any industry dependent on plant material. How is the latter processed and what are its uses?
6. Write brief notes on any four of the following:

   (*a*) Manihot
   (*b*) Cotton
   (*c*) Tea
   (*d*) Clove
   (*e*) Rubber
7. What are the major cereals grown in India? Mention the origin (native home) and botanical characteristics' of any one of them. What are the advantages of growing 'dwarf' varieties of cereals?
8. Give a general account of the plants whose economic products are exported from India.
9. Describe the uses and morphological nature of the plant parts yielding economic products in any *five* of the following:

   (a) Colchicum autumnale
   (b) Crocus sativus
   (c) Beta vulgaris
   (d) Elettaria cardamomum
   (e) Papaver somniferum
   (f) Glycyrrhiza glabra
   (g) Rauvolfia serpentina, and
   (h) Sesamum orientate

10. Describe in *any four* of the following the uses and the morphological nature of the plant yielding the economic products:
    (a) *Zingiber officinale*
    (b) *Saccharum officinarum*
    (c) *Syzygium aromaticum (Eugenia caryophyllata)*
    (d) *Ricinus communis*
    (e) *Cicer arietimum,* and
    (f) *Oryza sativa*
11. If you are banished to an uninhabited island 200 miles south of Ceylon and are permitted to take seeds of *only five* species with you to grow plants for your needs in the island, which species will you select and why? Write the botanical names and families of the species selected by you.
12. What is fibre? *Name four* fibres of commercial importance in India and write botanical names of the plants which yield them. In each case mention the family to which the species belongs, the plant part from which the fibre is obtained, and the economic uses of the fibre.
13. Give a brief account of any two of the drug plants you have studied with special references to their uses in pharmaceutical industry.
14. Describe the morphology, method of cultivation and processing of a beverage which earns foreign exchange.
15. Describe the uses and the morphological nature of the plant parts yielding economic products in *any five of* the following:
    (a) *Cinnamomum zeylanicum*
    (b) *Glycine max*
    (c) *Arachis hypegaea*
    (d) *Tectona grandis*
    (e) *Corchorus capsularis*
    (f) *Saccharum officinarum,* and
    (g) *Triticum aestivum*
16. What is a spice? Give a general account of any four important spices of India.
17. Describe the uses and morphological nature of the plant parts yielding economic products in any five of the following:
    (a) *Gossypium*
    (b) *Oryza*
    (c) *Cinchona*
    (d) *Pinus*
    (e) *Cicer*
    (f) *Papaver,* and
    (g) *Cocos*

18. What is meant by 'Green Revolution? Discuss the role of genetics in bringing about green revolution in India.
19. The Govt. of India has given top priority for the improvement of pulses in India in the coming years. What, in your view, are the objectives of the programme?
20. Write a short account of the processing of tea leaves. What do you mean by black tea and green tea?
21. *Write Short Notes:*
    (*a*) Wood seasoning
    (*b*) Economic importance of groundnut
    (*c*) Origin of wheat
    (*d*) Improvement of sugarcane
22. What is a fibre? Name three fibres of commercial importance that are grown in India. Write the botanical names of the plant which yield them, the family to which each belongs, the plant part from which the fibre is obtained and the economic use of the fibre.
23. Give a general account of the cultivation and economic importance of spices studied by you.
24. *Write Short Notes*:
    1. Green Revolution
    2. Rice cultivation
    3. *Cinchona*
    4. *Tectona*
    5. Green tea
25. What are the cereals? Give a brief account of wheat with special reference to its botany, methods of cultivation and improvement through breeding in India.
26. Describe morphology nature and uses of the plant parts is yielding economic products in any 5 of the following:
    (a) Arachis
    (b) Coffea
    (c) Corchorus
    (d) Curcuma
    (e) Gossy pium
    (f) Rauwolfia
27. Give botanical name, family and botany of the part used in any 5 spices studied by you.

Index

B

F

G

H

M

Q

R

S

X

Y

Z